Franck Kalala Kaniki
Paul Lunda Lubamba

The contribution of magnetometry to geological and structural mapping

Franck Kalala Kaniki
Paul Lunda Lubamba

The contribution of magnetometry to geological and structural mapping

The case of the KAPONDA sector in the DRC

Imprint

Any brand names and product names mentioned in this book are subject to trademark, brand or patent protection and are trademarks or registered trademarks of their respective holders. The use of brand names, product names, common names, trade names, product descriptions etc. even without a particular marking in this work is in no way to be construed to mean that such names may be regarded as unrestricted in respect of trademark and brand protection legislation and could thus be used by anyone.

Cover image: www.ingimage.com

This book is a translation from the original published under ISBN 978-620-6-71325-8.

Publisher:
Sciencia Scripts
is a trademark of
Dodo Books Indian Ocean Ltd. and OmniScriptum S.R.L publishing group

120 High Road, East Finchley, London, N2 9ED, United Kingdom
Str. Armeneasca 28/1, office 1, Chisinau MD-2012, Republic of Moldova, Europe
Printed at: see last page
ISBN: 978-620-7-77582-8

Copyright © Franck Kalala Kaniki, Paul Lunda Lubamba
Copyright © 2024 Dodo Books Indian Ocean Ltd. and OmniScriptum S.R.L publishing group

Epigraph

"The beginning of all the sciences is the astonishment that things are what they are.

Aristotle

In memoriam

A special thought to my late father François Kalala Kaniki Tshitata, to whom many qualifiers were attributed, and I can only quote his daily gift to me: "Tshena ni buanga bua ku kupesha to, the best I could offer you is the way of God and studies".

Kalala kaniki Franck

Dedication

I dedicate this work to: mom Monique Kapinga Kalala, dad Jean-Bosco Kadima, the big family of ten KALALA (Dior, Hélène, Ange, Patrick, Doryne, Paul, Peter, Franck, Horeb and Isaac), my study group Keep calm and do geology (Balega Muza- liwa Kennedy, Kabanda Mawanga Fiston, Luboro Mambonda Emmanuel, Lunda Lubamba Paul, Kimuni Mwamba Joseph, Kitenge ndjibu G.elias, Kabengana Mukala Eben and me), my pastor Patrick Kalenga, my friend Elie Stopack Mirindi , Patricia kele and all my colleagues.

Kalala kaniki Franck

To you, my parents Lubamba Ntambue Dimitrios and Bankate Ngoyi Délice,
To you, my brothers and sisters,
Friends, Colleagues and Acquaintances.
I dedicate this work to you

Lunda Lubamba Paul

Thank you

Our thanks go first and foremost to God our Creator, master of time and circumstances, for the breath he never ceases to renew in us.

our thanks also go to our Director, Professor Kadima Ka- bongo Étienne, and our co-director, Mr. Mulumba Mukala Jean-Luc, for the space they gave us in their occupations during the realization of this final Bachelor's project, and from them we learned the best way to exploit our potential.

To all our colleagues in the class of 2020-2021 and to all those who participated in one way or another in the realization of this work.

Table of contents

GENERAL INTRODUCTION

Fundamental geophysics and applied geophysics have often exchanged methods and results for the benefit of both, and what we present in this text has no other ambition than to continue these fruitful exchanges. Magnetometry is one of the main methods used in geophysics. It uses a magnetometer to measure the intensity of the Earth's magnetic field and/or its components. Although the magnetic method is one of the oldest methods in geophysics, it continues to attract a great deal of interest from its users. Indeed, the singularity linked to the exploitation of a natural field and the development of new techniques that enable large areas to be covered at judicious rates, make geophysical methods the most widely and transversally used according to the statistics of Telford et al. in 1976; Reynolds in 1997; Allard et al. in 1999; Reeves in 2005 cited by Kpirgbene [2016].

Geophysics applied to mineral prospecting allows us to tackle problems using three classic approaches Michel Allard [1999]: (1) the direct approach, which uses one or more geophysical methods to directly detect minerals in geological formations; (2) the indirect approach, as its name suggests, applies when direct detection proves impossible, and we then proceed by association; (3) the cartographic approach, which is much more cross-disciplinary and enables us to meet the contemporary challenges associated with mineral exploration. Conventionally, this two-dimensional approach is used to delineate geological contacts, locate structural elements and identify certain geological formations. The aim is to produce a less evasive geological map, or less evasive geological profiles, that can provide the best possible information on the nature and characteristics of geological formations, incorporating information on their physical properties. This information can be an indispensable tool for successful investigation of the earth's shallow, semi-deep or deep crust Kpirgbene [2016].

Since most of the earth's surface is covered by sediments and vegetation, it is important to exploit other techniques to map the geology through this cover. In the south-eastern part of the Kipushi mine, the local geology is still undifferentiated from the geological formations of the Katanga Supergroup. To solve this problem of undifferentiation between geological units, the application of the horizontal gradient to the magnetic field data usually enables the contacts between geological units to be determined. The aim of this thesis is to highlight the various contacts between geological units, including those of the Nguba and Kundelungu in our study area, located in the Kipushi territory in the Kaponda chiefdom (Haut-Katanga province).

Thus defined, our study is subdivided into 3 chapters apart from the introduction and conclusion.

— Chapter 1 provides a general overview of the geographical and geological context of the study area;

— Chapter 2 deals with the theoretical and methodological framework;

— Chapter 3 presents the various data processing and interpretation operations.

Chapter 1: GENERAL

I.1 Geographic setting

I.1.1 Location

Our study area is located in the Kaponda chiefdom, 5 km east of the town of Kipushi in Haut-Katanga. The geographical coordinates of this area in UTM zone 35S, datum WGS84 are between northing 8694500 m and 8699250 m on the one hand and between easting 531500 m and 540000 m on the other. Figure I.1 shows a map of the study area:

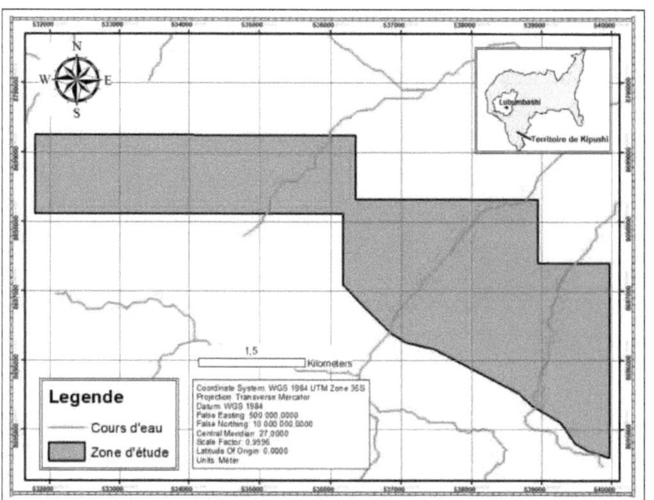

FIGURE I.1 - *Location map of the study area, with the Kipushi territory around the city of Lubumbashi in white on the map card.*

I.1.2 Hydrography

The hydrography of the Kaponda sector remains less well known, but that of the town of Kipushi is watered by several rivers and streams, including the 135 km-long Kafubu River, which rises in the village of Shimpauka in the Inakiluba groupement (Kaponda chiefdom) and crosses the territory from east to west, flowing into the Luapula River at the village of Kanga in the Kiniama chiefdom (one of the territory's tourist sites). The main rivers are : Bwishibila, Munama, Musoshi, Kafubu, Kifumanshi, Kiswishi and Luapula Patient [2020]. Figure I.2 shows the hydrography of the study area:

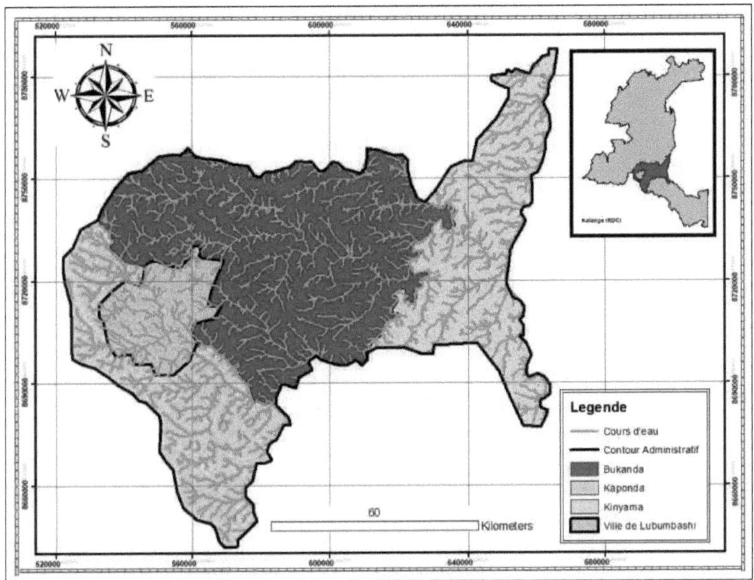

FIGURE I.2 - *Hydrographic map of Kipushi, the mapboard shows Haut-Katanga province and the Kipushi territory in red.*

I.1.3   Climate and vegetation

Classified as a humid temperate climate with a dry season, Kipushi enjoys a tropical climate. Average annual rainfall over the past 15 years has been 1,260 mm Patient [2020]. The average annual temperature is around 19.8°C. The vegetation found in the Kipushi territory is generally open forest; the plant cover is predominantly populated by species from the grass and legume families. The Kipushi territory forms the green belt of the city of Lubumbashi. The forest is a particular feature of the area, which contains a large part of the *Miombo* open forest. Investment in rational exploitation, which would ensure that the ecological balance is maintained, would be a very important element in the development of the territory, based on forest resources, the main source of several products for generating energy (charcoal and firewood), healthcare (medicinal plants), food (non-wood forest products: fruit, mushrooms, honey, etc.); and woody species used in carpentry and construction. Patient [2020]

I.1.4   Geomorphology

The geomorphological map of the city of Kipushi and Lubumbashi is shown in Figure I.3 in 3D.

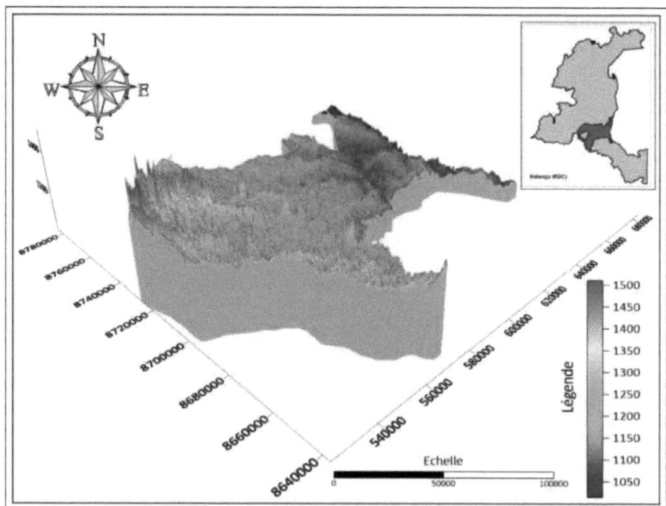

FIGURE 1.3 - *Map of the geomorphology of Kipushi and Lubumbashi, the mapboard shows Haut-Katanga province and the Kipushi territory in red.*

The topography of the town of Kipushi is characterized by hills, some of which reach 1,350 to 1,400 m in altitude, and plains occupied by villages and former Gécamines workers' camps. The town of Kipushi lies on a plateau dominated by closely-spaced lapis, separated by sharp rocky snags due to the karstic topography. The valleys are occupied by streams, some of which are seasonal. This area is part of the Katanga Bilolo [2016] high plateau geomorphological complex.

I.2   Geological setting

1.2.1   Regional geology

According to François [1987a], the greater Katanga area shows geological formations divided into two groups: bedrock formations of Proterozoic age and overburden formations of Phanerozoic age.

The Proterozoic formations are folded and divided into three major groups, including :

⇒ The Ubendian chain of Paleoproterozoic age (1700Ma);

⇒ The Kibaran chain or Kibara Supergroup of Mesoproterozoic age (1450900Ma) (Kokonyangi [2004]) ;

⇒ The Katanguian chain or Katanga Supergroup of Neoproterozoic age (1000530Ma) (J. Cailteux [2019]).

The Phanerozoic age formations are sub-tabular and represent the sedimentary cover. From bottom to top, they are made up of :

⇒ The Biano Group of Paleozoic (Cambrian) age (J. Cailteux [2019]) ;

⇒ The Karoo series of Paleozoic age (Permo-Carboniferous) (D. Delvaux [2010]) ;

⇒ The Kalahari series of Cenozoic age;

⇒ Recent alluvium and overburden of Quaternary age.

The Kipushi Territory is underlain by Neoproterozoic formations represented by the Katanguienne Range or Katanga Supergroup.

1.2.2   The Katanga Supergroup

The Katanguien is made up of Neoproterozoic to Lower Paleozoic formations, part of which was affected by folding that led to the formation of what is known as the "Lufilian Arc", with a NW-SE direction of elongation (training), its convexity facing the Septentrion (the North) and the other part forming the foreland of the arc, known as the "Tabular Katanguien". It is underlain by crystalline formations of Ubendian age and Mesoproterozoic formations of the Kibaran Range (François [1987a], J. Cailteux [2005],

J. Cailteux [2007]).

The Tabular Katanguien is bounded to the northeast by the Bangweulu Block, composed of Archean to Paleoproterozoic granitoids and volcanic rocks and the Paleoproterozoic Ubendienne Range, to the southeast by the Mesoproterozoic Irumide Range and to the west by the Kibaran Range. It rests unconformably on the Kibaran Range to the north and northwest (Cahen [1970], J. Batumike [2008] cited by Kipata [2013]) and is thought to have been deposited in a rift extension context. The total thickness of the Katanga Supergroup varies between 7 and 10 km (Kampunzu [1998], J. Cailteux [2019]).

Litho-stratigraphy, geodynamic and paleogeographic frameworks of de- pot environments

The Katangan chain extends from Zambia to the south-east of the DRC, i.e. Katanga in general. Using two tillite levels (Diamictites) as landmarks (the Grand and Petit Conglomerates), it has been subdivided into three groups, whose regional importance remains undeniable. (François [1973], François [1987b], J.Batumike [2007], J. Cailteux [2005]). From bottom to top, they are as follows:

⇒ The Roan group (separated from the underlying group by the Grand

Conglomérat), ⇒ The Nguba group (separated from the underlying group by the

Petit Conglomérat), ⇒ And the Kundelungu group.

The litho-stratigraphic synthesis of the Katanguian is shown in figure I.4.

Supergroup	Group	Subgroup	Training	Lithology
KATANGUIEN	KUNDELUNGU: ± 500 Ma	Biano Ku3		Arkoses, Conglomerates and Clay-sandstones
		Ngule Ku2	Sampwe Ku2.3	Dolomitic pelites, Clay-micrograin
			Kiubo Ku2.2	Dolomitic sandstones, microgres and pelites
			Mongwe Ku2.1	Dolomitic pelites, microgres and sandstones
		Gombela Kul	Lubudi Kul.4	Alternating limestone and sandy-carbonate beds
			Kanianga Kul.3	Shales and carbonate microgres

Age	Formation	Unit	Description
		Lusele Ku 1.2	Pink limestone
		Kyandamu KuLl	Small Conglomerate (tillite/diamectite)
NGUBA: 620Ma	Bunkeya Ng2	Monwezi Ng2.2	Dolomitic sandstones, microgres and pelites
		Katete Ng2.1	Dolomitic sandstones alternating with shales
	Muombe Ngl	Kipushi Ngl.4	Dolomites and dolomitic shales
		Kakontwe Ngl.3	Carbonates
		Kaponda Ngl.2	Carbonate shales and microgres
		Mwale Ngl. 1	Grand Comglomerat (tillite / diamectite)
ROAN : 750 Ma	Mwashya R4	Kanzadi R4.3	Sandstone or alternating Shales and Microgres
		Kafubu R4.2	Carbonate shales
		Kamoya R4.1	Dolomitic shales, Microgres, Sandstones, Beds with comglomeratic inclusions and Cherts
	Dipeta R3	Kansuki R3.4	Dolomites with volcanic interlayers
		Mofya R3.3	Dolomites, dolomitic microgres and arenites
		Fungurume R3.2	Clayey-dolomitic micro-gravel with intercalation of feldspathic sandstones or Dolomites
		R.G.S. R3.1	Clayey-dolomitic micro-gravel (sandstone-schist)
	R2 mines	Kambove R2.3	Laminated clay-limestone dolomites and stroomatolites
		S.D R2.2	Dolomitic shales with 3 carbon horizons
		Kamoto R2.1	Stromatolitic dolomites (RSC), silicified and arenitic dolomites (RSF/D.Strat) and clay-dolomitic microgres (gray RAT)
	RAT Rl		Red clayey-dolomitic microgres and sandstones " Clayey-talcic rocks<<
The basis of RAT is not			ɔien known
<900Ma		**Comglomerate base**	

FIGURE I.4 - *Iithostratigraphic synthesis of the Katanguian (François, 1973; Cailteux et al., 1994; Batumike et al. Kipata, 2013; modified by Cailteux and De Putter, 2019; Kanzundu, 2020)*

I.2.2.. 1 The Roan Group

The Roan Group is largely made up of dolomites and do- Iomitic shales (Cailteux [1981], J. Cailteux [2007]). It is subdivided into four subgroups from bottom to top: R.A.T (Roches Argilo-Talqueuse) " Musonoi actuellement" *(R.1),* Mines *(R.*2), Dipeta "Fungurume actuellement" *(R.3)* and Mwashya(*R.*4). Two major cycles of transgression and regression are recognized (Cailteux [1981], J. Cailteux [2019]). Each cycle begins with continental to sub-continental deposition (*R.* 1 and *R.*3) and ends with carbonate platform sediments (*R.*2 and *R.*4).

The R.A. subgroup T(*R.*1)

These are soft formations made up of red clayey-dolomitic micro-sandstones and sandstones. These rocks are oxidized and affected by many detachments. This is the reason for their discontinuity in the field (J. Cailteux [2005]). The basis of this subgroup (present-day Musonoi) is still unknown.

The Mines sub-group *(R.2)*

II comprises three formations, from bottom to top, including: the Kamoto formation (*R.*2.1); the S.D (*R.*2.2) and Kambove (*R.*2.3). The base of this subgroup consists of stromatolithic dolomites that denote reducing conditions. The dolomites of this subgroup are related to platform carbonate deposits in tides and lagoons with reducing conditions (J. Cailteux [2005], AB Kampunzu [2005]).

The Dipeta subgroup (*R.*3)

Previously, Dipeta (now Fungurume) was subdivided into four formations from bottom to top: R.G.S (*R.*3.1) ;*R.*3.2 ; Mofya (*R.*3.3) and Kansuki (*R.*3.4) (J.Batumike [2007]), is currently five formations. These formations are: the Kwatebala formation at the base overlain by the Dipeta, Tenke, Mofya and Kansuki formations at the top (J. Cailteux [2019]).

This base is represented by oxidized clay-sandstone rocks, while the top is occupied by clay rocks of the Iagunaire type, which in the past were attached to the RGS of the Sebkha

Cailteux type [1981].

The Mwashya sub-group *(R.4)*

It comprises three formations: the Kamoya *(R.4.1)* at the base, the Kafubu *(R.4.2)* in the middle and the Kanzadi (*R.4.3*) at the top. These are sedimentary sequences made up of dolomitic shales, micro-sandstones, sandstones, cherts and a Conglomerate at the base, followed by Black shales and finally Sandstones and alternating shales and micro-sandstones (J. Batumike [2007], J. Cailteux [2005].
The base of the Kamoya formation (Mwashya Conglomerate) corresponds to a diamictite deposited during the Kaigas glaciation (770-735 Ma) (J. Cailteux [2019]).

I.2.2.. 2 The Nguba Group

Nguba is subdivided into two subgroups, with Muombe *(Ng.1)* at the base and Bunkeya *(Ng.*2) at the top (J.Batumike [2007]).

The Muombe subgroup *(Ng.1)*

It begins with the Mwale Formation *(Ng.1.1)* consisting of a diamictite called "Grand Conglomérat" (AB Kampunzu [2005]), followed by the Kaponda (*Ng.*1.2), Kakontwe (*Ng.* 1.3), Kipushi *(N g.1.4)* (J.Batumike [2007], Cailteux et al., 2007). Currently, a new carbonate formation "Dolomie Tigrée" is being introduced above the Mwale Formation, which is now considered a separate formation from the Muombe Subgroup (J. Cailteux [2019]; Delpomdor et al., 2020).

Bunkeya subgroup (*Ng.*2)

Consists of the Katete formation (*Ng.*2.1) and the Monwezi formation (*Ng.*2.2), mainly dolomitic sandstones, microstones and shales.
In the southern part of the arc, the Katete Formation is represented by an alternation of dolomites and shales long known as the Recurrent Series (Tshilolo, 1987; Batumike et al., 2007; Cailteux et al., 2007; Heijlen et al., 2008).

I.2.2.. 3 The Kundelupgu Group

The age of the Kundelungu formations ranges from around 620 Ma to around 502 Ma. It comprises the Gombela subgroup *(Ku.1)* at the base, the Ngule *(Ku.2) in* the middle and the Biano (*Ku.*3) at the top, according to the former subdivision by Batumike et al. (2007). At the base of the Kundelungu Group is the small conglomerate (diamictite with a purple-grey clay-carbonate matrix and abundant cm elements), the Kyandamu Formation, which is currently considered separate from the Gombela Subgroup. The Gombela Subgroup then comprises the Lusele (*Ku.*1.2), Ka- nianga (*Ku.*1.3) and Lubudi (*Ku.*1.4) formations. The Ngule subgroup comprises the Mongwe (*Ku.*2.1), Kiubo (*Ku.*2.2) and Sampwe (*Ku.*2.3) formations. The Mongwe and Kiubo formations are essentially detrital with variable carbonate content. Taken together, they yield a thick layer that is interpreted as a molassic deposit that filled the Kundelungu basin just after the first phase *(D1)* of folding towards the end of deposition of the Gombela subgroup (François, 1973; Cailleux and De Putter, 2019; Ilunga and Cailteux, 2019). The top of the Kiubo Formation is occupied by a major tectonic unconformity that represents the basal part of the nappes that were thrust; locally we find mega breccias forming the base of these nappes (Cailteux et al., 2018). The Biano subgroup is made up of arkoses, conglomerates and clayey sandstones and is thought to have been deposited during the Cambrian. Currently, this subgroup is considered a separate group from the Kundelungu group (Batumike et al., 2007; Cailteux and De Putter, 2019; Ilunga and Cailteux, 2019).

I.2.2.   a Some modifications to the stratigraphy of the Katanga Supergroup

Changes in nomenclature and definition of formations, subgroups and groups have been proposed for some time. Much more recent investigations have just revised the lithostratigraphy (Cailteux and De Putter, 2019), nomenclatures and definitions proposed by (Batumike et al., 2007; Cailleux et al., 2007). Currently, the R.A.T subgroup is renamed Musonoi while including the lower limb R.A.T grise in this Musonoi group. The Dipeta subgroup is renamed Fungurume with a change and new definition in the formations, the R.S.G formation becomes Kwatebala, the *R33.2* formation is split into two formations including Dipeta and Tenke. The Mofya formation (R3.3) remains with the Mofya nomenclature. The Mwale and Kyandamu formations are isolated from the Muombe and Gombela subgroups respectively, and treated as formations in their own right. A carbonate cap, the Dolomie Tigrée, is introduced above the Mwale formation and constitutes a new formation.

The Biano subgroup is redefined as a separate group detached from the Kunde-Lungu with a Cambrian age. A major tectonic unconformity is recognized towards the base of Musonoï, another at the base of the Kansuki formation. Between the Kiubo and Sampwe formations, a new tectonic unconformity appears, and between the Kundelungu and Biano groups, a stratigraphic unconformity. The existence of thick carbonate layers or Cap carbonates (Muombe and Gombela) above diamictites/tillites (Petit and Grand Conglomérat) supports the hypothesis of a totally glaciated Earth, where continents are covered by ice caps and oceans by pack ice "Snow Ball Arth Theory" (Hoffman et al., 1998; Hyde et al., 2000, Kipata [2013] et al., 2013, Cailleux and De Putter, 2019; Delpomdor et al., 2020).

I.2.2. b Tectonics

During the Lufilian orogeny, part of the Pan-African Orogenesis, the Katanga Supergroup formations were folded and thrust. This orogeny gave the sediments of the Katanga Supergroup their current arc-like configuration. Quoted in Batumike (2008), Binda and Porada (1995) established the tectonic zonation that divides the Katanga Range into five domains:

⇒ External fold-thrust belt;

⇒ The domes area ;

⇒ Synclinariale belt;

⇒ Katanga high;

⇒ The Katanga aulacogen.

Demesmaeker et al , (1970) and François [1987a] distinguish three sectors with different tectonic effects in the "External fault and trust belt" area, including :

1. The south-east sector: tectonics here are characterized by complex anticlines

2. Central sector: the tectonics are extrusive and the folds dip southwards. These are the Likasi, Shinkolobwe, Kambove and Fungurume regions.

3. The western sector: tectonics here are also extrusive, overlapping and end in thrusting. The Kolwezi sector has a highly complex faulted structure.

According to Kampunzu and Cailteux (1999), the Lufilian orogeny appears to have operated in three phases that contributed to the shaping of these sediments. These are :

⇒ The *D1* phase (Kolwezian phase) was characterized by folding and thrusting, with the structures being transported in a northerly direction. The core of the anticlines was often affected by faults and occupied by tectonic breccias. It is thought to have taken place between 800 and 710 Ma, reaching its climax between 790 and 750 Ma.

⇒ Phase D2 (Monwezian phase), which affected the folded and overlapped terrains by E-W-trending senestial strike-slip faults in the western part of the belt (Monwezian fault system). This phase appeared between 690 and 540 Ma and can be correlated with the orogeny that affected the Dammarian Range to the SW;

⇒ D*3* phase (Chilatembo phase), it is considered responsible for the straight, open NE-SW trending folds orthogonal to the arc and the conjugate N160-N170E and N70-80E trending folds in the eastern part of the belt, suggesting NW-SE trending compression (example of the Chilatembo syncline). The agent of this event is not yet well known. However, it is thought to be Lower Paleozoic, and therefore inferior to 540 Ma.

Further development of this tectonic model with kinematic fault and paleo stress inversion analysis demonstrates a total of eight stages of brittle deformation (Kipata [2013]; Kipata et al., 2013). Of these eight stages, five are related to Lufilian Orogenesis and three are post-orogenic. The eight deformation stages are:

⇒ The first stage: Early N-S-trending compression, expressed by unmineralized fractures. Structures have a preferential E-W direction.

⇒ The second stage: Constriction in the central part of the outer Lufilian Arc, accompanied by tectonic breccia injections with intense block rotations in a context of vertical extrusion. It is attributed to constriction-type deformation caused by the monoclinal curvature of the Lu- filian Arc and considered to be synchronized with rincurvation itself.

⇒ The third stage: NE-SW regional transgression, well expressed in the Arc foreland and on the margin of the Kibaran chain. This stage is correlated with the NE-SW compression observed in the Ubendian chain.

⇒ The fourth stage: Transtension after a permutation of σ_1 and σ_3 stresses. It developed an almost orthogonal extension of the Arc in the central and southern part, it is probably the result of a relaxation after a NE-SW shortening of the third stage. It marks the start of a late-orogenic extension in the Lufilian Arc.

⇒ The fifth stage: Extension parallel to the Arc, it is characterized by normal faults with mineral remobilizations (malachite impregnations are found in striations on some fault planes), and marks a late-orogenic extension.

⇒ The sixth stage: Reversal of the NW-SE regional transgression, this stage consists of unmineralized fractures and would have developed mesoscopic folds orthogonal to the preferential direction of the outer Lufilian Arc in its SE part. Luiswishi and Mine de l'Etoile cases.

⇒ Seventh and eighth stages: a NE-SW and NW-SE extension, these two stages are linked to an extensional context that would have prevailed in East Africa after a reversal of the transgression.

Other work shows the correlation between the three phases established by Kampunzu and Cailteux (1999) and some of Kipata's brittle deformation stages (Kipata [2013]; Kipata et al., 2013) is as follows:

⇒ The *D1* phase is linked to the first stage, the first stage of deformation characterized by thrust faults, the characteristic faults of which are essentially unmineralized; the curvature of the Arc corresponds to the second stage and is always linked to the *D1* phase. This phenomenon is known as "ben ding". It has been interpreted as the probable result of compression controlled by lateral stresses generated by the Kibaran chain to the NW and the Bangweulu block to the east;

⇒ Phase D2 corresponds to the third stage, where the deformation regime is unstuck with transgressive deformation characterized by unstuck faults of regional extension ;

⇒ Phase *D3* corresponds to the sixth stage, which would have occurred after the fourth and fifth stages, both of which could reflect gravitational flow and late to post-orogenic extension. Phase *D3* is therefore postorogenic, a Permo-Triassic inversion in transpressive rifting found in other parts of Africa.

The Nguba and Kundelungu, as well as the Mwashya subgroup, are linked to a very large phase of extension and normal faulting that marks the transition from rifting to a proto-Ocean of the Red Sea type (Buffard, 1988; Cailteux et al., 2007).

I.2.2. c Magmatism

Magmatism in the neoproterozoic formations of the Katanga Supergroup is represented by :

⇒ Basic rock cinerites in the formations of the Sub-Groupe des Mines du gisement de l'Etoile, in the Kambove sector and in the Luishya mining polygon (Lefebvre and Cailleux, 1975) ;

⇒ Sills and dykes of gabbroic and dioritic rocks in the bedrock of the Fungurume Subgroup (ex-Dipeta) in the Kakonge, Mwadingusha, Makawe, Shinkolobwe and Kipushi sectors (Lefebvre and Cailteux, 1975; Batumike et al., 2007; Mashala,

2007) ;

⇒ Basic pyroclastites found under variable aspects (tuffs, lapillis, argillites) in the Kansuki formation (formerly Lower Mwashya) (Lefebvre, 1973; Cailleux et al., 2007; Cailleux and De Putter, 2019) ;

⇒ Basic and dioritic lavas at the base of the Grand Conglomérat in the Kibambale region near Mitwaba and Basalts at Kasenga. Kimberlite pipes recognized in the Kundelungu plateau (Batumike, 2008; Batumike et al., 2008).

1.2.2. d Metamorphism

According to François [1987a], the Katanga Supergroup underwent metamorphism that increased in intensity from north to south and from east to west. This metamorphism resulted in the transformation of the clay minerals of the original sediment into authigenic sericite and chlorite. Sericite is more abundant than chlorite, except in a few Roan horizons. It is characterized by the occasional presence of neoformational albite, as well as by more or less complete talcification of certain tetanized dolomite beds. Unrug (1983) distinguishes three metamorphic zones from Zambia to Katanga. These are :

⇒ The sericite and chlorite zone from Lubumbashi to Kengere towards the north of the Katanguien basin;

⇒ The scapolite-epidote-actinote zone, in this zone the metamorphism is of the amphibolite facies type as we note the presence of mineralogical assemblages including Musoshi disthene, Kitwe to Lambo-Kisinga ;

⇒ The Amphibole-Grenat zone from Kisinga to Solwezi.the metamorphic assemblage represented by chlorite-biotite-disthene indicates that temperatures may have varied between 4000°C and 6000°C under pressure conditions ranging from 2 to 8 Kb. This shows that the Katanga Supergroup formations underwent Barrow-type metamorphism.

1.2.2. e Mineralization in the Lufilian arc

The main mineralizations known in the Lufilian arc of Katanga with regard to their lithostratigraphic position are :

Nguba-related Zn-Cu-Pb mineralization, (Intiomale and Oosterbosch [1992] ;Intiomale [1982] ; Chabu.M [1990]. This type of mineralization has been associated with polymetallic vein deposits,

Brown [1979] distributed along major faults that developed during the Lufilian orogeny. Intiomale [1982] and Oosterbosch [1992]; (Kampunzu et al., 1998) interpret these deposits as formed in compressive regimes associated with major thrusting marking the closure of the Katanguien Basin.

Iron deposits are mainly located in the Lower Mwashya in southern Katanga (see Oosterbosch [1992]; François [1987a] and Cailteux [1981]). These are stratiform deposits in which the ore is expressed as magnetite, oligite or goethite and appears in massive or banded beds. These occurrences have therefore been classified by François [1987a] and Cailteux [1981] as itabitic formations. In the Nguba strata, as in the Kisonga deposit near Kambove, these deposits are in the form of clusters. The iron-bearing rocks that appear near Kengere and to the south of the Lufilien Arc in Katanga are located in the undifferentiated Roan (François [1987a] and Cailteux [1981]).

The Upper Mwashya around Lubumbashi and Mulungwishi is abundantly affected by hematite veins. The Moa-Mululu deposit on the Lukafu road and the Kanunka deposit between Mulungwishi and Luambo, formerly attributed to the Sous-Groupe des Mines, are now associated with the Upper Mwashya. Iron deposits are also observed in the detrital Lower Roan of the SE portion of the Lufilian Arc (Bala-Bala, 1985), in the lilac RAT of the base of the Katanguien and also in the Kundelungu shales, where the ore appears in the form of oligite flecks and veinlets (Cahen [1977]).

Copper-cobalt mineralization is by far the most important economically. They are classically hosted in the Mining Subgroup and exceptionally in the Lower Mwashya at Shituru and Tilwezembe for the NW copper sector, but also in the Nguba at Kipushi, Lombe and Kengere. In the SE district of Katanga and Zambia, these mineralizations are hosted in the coarse Roanian detrital formations of Mutanda (Kisenda and Lubembe mines) Ngoy(1995) and ore-shales at Musoshi (Tshiauka et aL 1995). They are rigorously stratiform in the Mining Subgroup, where they constitute two mineralized bodies, lower in the Kamoto dolomite (D. Strat. and RSF) and upper at the base of the Dolomitic Shales, as well as lenticular bodies in the Kambove dolomite or CMN (Calcaire à minéraux noirs) (Cailteux, 1994).

Discussing the evolution of genetic patterns in the copper-cobalt deposits of the Katanga and Zambian Copper Arc (Okitaudji, 1997), following Cailteux [1981], demonstrated that the concentrations of these metals were syngenetic. The metals present in the sediments are trapped by the sulfur produced by the bacterial reduction of marine sulfates. This process leads to stratiform, bedded orebodies that may be enriched or depleted in places as a result of later remobilization of metamorphogenic fluids from before the structuring of the Groupe des Mines and later supergene remobilization.

I.2.3   Local geology

Figure I.5 illustrates the local geology of the study area:

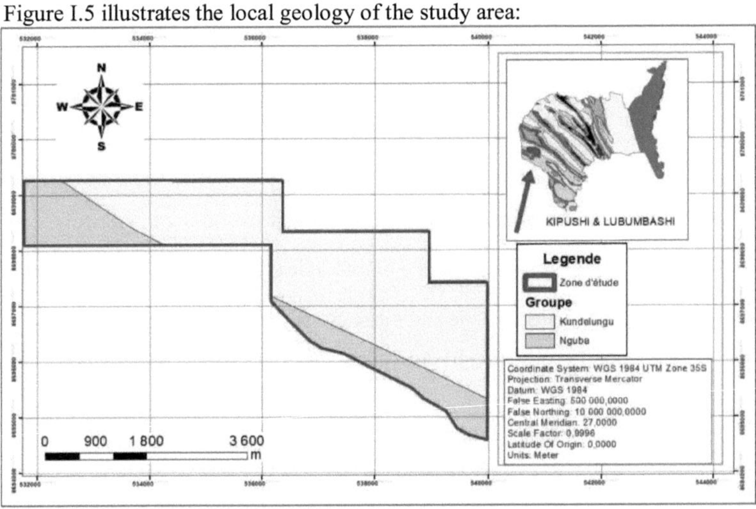

FIGURE I.5 - *Local geological map of the study area, the mapboard shows the regional geology of the Kipushi territory and the city of Lubumbashi, the red arrow indicates our study area in red.*

Our study area is underlain by Neoproterozoic bedrock, as can be seen on the previous map, with only the two Katanguian groups outcropping: the Nguba and the Kundelungu. The geological formations are oriented north-west-south-east and dip north-east, as shown by Intiomale [1982] and Nyembo [1997].

I.2.3.. 1 Litho-stratigraphy

The Katanguian lithostratigraphy modified by François [1987a], (Batumike et al., 2007 and Cailteux et al., 2007), Kipata [2013], as well as Kanzundu [2020] shows the following lithostratigraphy in our study area:

1.2.3. a The Nguba

In this group, the following training courses are presented:
⇒ of Kipushi (Ng 1.4) consisting of dolomites and dolomitic shales,
⇒ of Katete (Ng 2.1) consisting of shales alternating with dolomite beds called the "Recurrent Series",

⇒ and Monwezi (Ng 2 ;2) consisting of dolomitic Sandstone and Microgres.
These formations are undifferentiated (there are no geological contacts between them).

1.2.3. b Kundelungu

In this group, the following formations are present According to François

[1987a]: ⇒ Lusele (Ku 1.2) consisting of pink micritic dolomites, ⇒ Kanianga (Ku

1.3) consisting of carbonate shale and microgres,
⇒ Mongwe (Ku 2.1) consisting of dolomitic pelites, micro-sandstones and sandstones;

⇒ Kiubo (Ku 2.2) consists of dolomitic sandstone, micro-sandstone and pelite.
It should be noted that these Formations also remain undifferentiated.
Earlier work shows the NW-SE Kaponda level (Ng 1.3), 111.30 m thick at Kipushi, of which 76.70 m is essentially a blue-grey dolomite with alternating light and dark laminations known as tiger dolomite (François, 1973).

Hydrothermal deposits in stress zones are found in areas of uneven pressure, where networks of discontinuous fissures sub-parallel to the compressed horizons have developed, generating hydrothermal circulation capable of producing valuable ores.

One of these is the Kisanga type, consisting of lenticular or elliptical, but thick, pyrite substitution clusters with Zn, Cu, Pb, As, Co occurrences, in laminated dolomites subjected to strong torsion at the Kaponda-Kakontwe contact. The ores alter into hematite-limonite containing various trace elements.

I.2.3. c Structure

The Katangan orogeny observed in southern Katanga determines a morphological detail from the Congo-Zambia border towards the Rwashi, materialized by the following succession: Kipushi anticline, Kafubu syncline, Lupoto anticline, Lubumbashi syncline and Rwashi anticline. The Kipushi Anticline is a slightly *NNE-dipping* fold whose northern flank dips between 75° and *85°NE* and whose southern flank dips between 60° and 70° southwards Intiomale [1982]. The Kipushi polymetallic copper-lead-zinc deposit, a vein-shaped concentration, is located at the NW end of the anticline's northern flank. The Katangan orogeny produced two main tectonic faults, listed below:

1. **Kipushi Fault:** a steeply dipping transverse fault at around *70°NW* that cuts the Mwashya formations, the Likasi-Lubumbashi groups and the recurrent series on the northern flank. The fault is mineralized.

2. **the axial fault:** this is a longitudinal fault running *ESE-WNW,* filled by the Roan axial breccia. In the extension of the axial fault and with the Kipushi fault, a collapse zone filled with breccias can be seen Nyembo [1997].

Chapter 2: THEORETICAL AND METHODOLOGICAL FRAMEWORK

II.1 Theoretical framework of the magnetic method

Reproducing a precise image of everything hidden underground is apparently difficult and often unacceptable at first glance, yet the instruments (probes) used in medicine enable direct, efficient diagnosis and give a faithful picture of the state of the organ affected, enabling the doctor to immediately prescribe the right medication for the patient. All diagnoses carried out on the patient are either carried out in real time (as in the case of expert systems) or at a slightly later stage (it may take several sessions before the disease is localized). The decision to progressively treat the disease or remove the affected organ depends on the interpretation of the treating physician).

The highlighted terms used in this first paragraph have a similar meaning in geophysical subsurface exploration H.Shout [2004].

The magnetic method uses a probe (a magnetometer) to measure a physico-chemical property of matter (petrophysical parameter), namely magnetic susceptibility. In short, it involves studying the earth's magnetic field measured at the surface to establish the distribution of magnetic susceptibility in the subsurface.

II.1.1 The magnetic field

At any point on the earth's surface, the compass needle orients itself, and this orientation proves that a natural magnetic field linked to the earth exists. A study of the reciprocal action of a magnet and a magnetized needle reveals that both carry positive and negative magnetic masses that can be measured qualitatively.

Two magnetic masses m_1 and m_2 attract each other if they have opposite signs, and repel each other if they have the same sign, by forces proportional(F) to the product of their masses and in inverse proportion to the square of their distance r (C. Coulomb, 1736-1906) quoted by H.Shout [2004].

$$F = \frac{1}{\mu} * \frac{m_1.m_2}{r^2}$$

with: μ: the permeability of the medium ($\mu=1$ in a vacuum and in air).

The magnetic field at a point P is represented by the force that would act on a unitary magnetic mass m_1 $(m_1 = 1)$ located at this point. This force can be attractive if m_1 and m_2 are of opposite sign, or repulsive if m_1 and m_2 are of the same sign. The magnetic field is characterized by :

1. a point of application,

2. direction,

3. meaning,

4. a value (its intensity).

II.1.1.. 1 Magnetic field measurement units

The unit of the magnetic field is one of the following: In the CGS system we have the oersted *(Oe)* and the gamma(γ).

The oersted is the unit that measures the magnetic field intensity generated by a magnetic mass equal to unity *(m_1 = 1)*, placed at a distance of one centimeter from a given mass. In the CGS system, the dimension of œrsted is *cm1/2g1/2s-1*.

Since $\mu = 1$ in air (or vacuum), one Œrsted is equivalent to one gauss (defined as one maxwell per square centimeter *(M /cm$_x^2$)*. Since the Gauss is a unit of induction

11.1.2   The magnetic pole

The magnetic pole is by definition the point on the Earth's surface where a compass needle would point downwards, i.e. where the magnetic field is vertical. The North Magnetic Pole has a weaker magnetic field than the South Magnetic Pole, which means that the center of the dipole is slightly further away from the center of the dipole. The axis of this dipole intersects the earth's surface at 78.5 N and 69 W. Moreover, paleomagnetic studies show that this axis has never deviated too far from the axis of $11°5$ with the Earth's axis of rotation.

By analogy with electric charges as encountered in electrostatics, the notion of magnetic mass is defined in the same way in magnetism (it has no physical reality and is only used for formulation and calculations).

- $\Rightarrow$ Magnetic masses are assumed to be positive if they refer to the North Pole, in the case of a magnet or magnetic North ;
- $\Rightarrow$ Magnetic masses are assumed to be negative if they refer to the South Pole (South Magnetic Pole).

In the past, gamma(γ) was used to express the value of the field.

In the SI; the Weber *(Ж0)* has dimension: $Kgm^2\ s^2\ A^{-1}$.

At present, the unit used in magnetic prospecting is the nanotesla *(nT), whose* SI dimension is Kgs^{-2} *A-1*, with a transformation set exactly equal to the old unit, the gamma(γ). In Katanga, the total field strength *(F)* is over *31*,000nT.

$$(1nT) - 10^{-9}T - 1(\gamma) - 10^{-5}(Gauss) - 10^{-5}(Oe) - 10^{-9}(Wb/m^2)$$

11.1.3   The magnetic moment

There is no such thing as free magnetic mass; every magnetic body always has two opposite magnetic poles (south and north) defined in relation to the earth's geocentric pole. The magnetic moment(M) is directly proportional to the magnetic mass located at the magnet pole(Tm) and to the distance separating the magnet poles (l).

$$M = \pm m.l$$

On a microscopic scale, when protons and their electrons vibrate and rotate on themselves, they give the atom a proper magnetic moment, and the electrons' movement around the nucleus also gives them an angular magnetic moment. The magnetic moment of an electron is the sum of its intrinsic magnetic moment and its angular momentum.

II.1.4 The magnetic dipole

The magnetic dipole is an association of two magnetic masses $-m_1$ and $+m_2$ located at an infinitesimal distance d_l from each other. Usually we use the notion of dipole, which would be closer to the reality of a magnet materialized by two poles, we can associate a magnetic moment defined by the expression :

$$d_M = m.d_l$$

where d_l is the vector oriented from $-m_1$ to $+m$.

II.1.5 Magnetic susceptibility of rocks

Magnetic susceptibility is the ability of a rock to magnetize under the action of an excitation or ambient magnetic field.

$$K = \frac{T}{\beta}$$

where : T: being the magnetization induced by the β field.

The direction and intensity of magnetization depend essentially on the constant K, known as the body's magnetic susceptibility (magnetization coefficient).

II.1.5..1 Magnetization intensity

The magnetization intensity of a body represents the magnetic moment of a unit volume of this body or of an elementary volume d_v . it is defined by a magnetization intensity Jet has a magnetic moment d_M , such that :

$$I - \frac{d_M}{d_V}$$

II.1.6 Earth's magnetic field

Figure II.1 shows the components of the magnetic field:

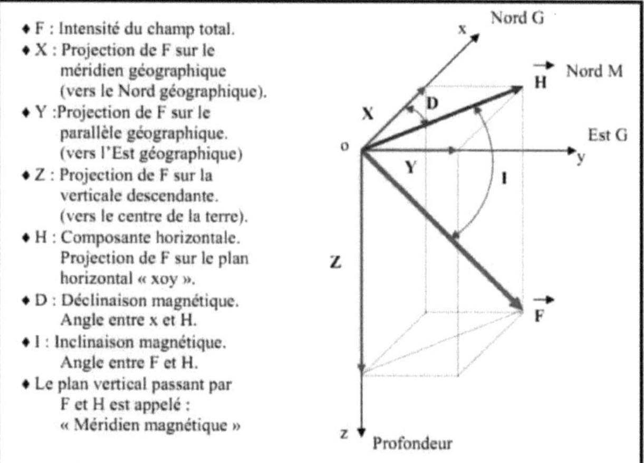

FIGURE II.1 - *Component of the Earth's magnetic field; Telford-1998*

The magnetic field can be defined by three components at any given point: north, south and vertical (x, y, z).

Very often, we give a value expressed by the magnitude of the total field F and its declination D as well as its inclination I; where D is the angle between the horizontal component of the field and geographic north and I, the angle between F and the horizontal Chouteau [2002]. The earth's magnetic field can be approximated by a dipole field. It is too complex to be expressed by a simple mathematical function, but it can be considered uniform over a few hundred km, and the geological background noise is easily observable.

F has an intensity of 0.6 Oe at the magnetic poles ($I=\pm0°$) and a minimum of 0.3 Oe at the magnetic equator ($I=0°$).

II.1.6.. 1 Main field origin

The main magnetic field can theoretically be caused by an internal or external source, whose change can be persistent or recorded by a current flow. Mathematical analyses of the field observed at the earth's surface show that at least 99% is caused by internal sources and 1% by sources external to the earth. Today's magnetic field is the result of three components: internal sources (main field), external sources (transient field), and induced sources (anomalous field).

II.1.6. a Internal magnetic field

The internal magnetic field is the main part of the field and corresponds roughly to that of a dipole (internal origin). This field varies slowly over time, particularly over the centuries, in what is known as "secular variation". Several theories have been put forward to explain the mechanisms of internal sources, the most explicit being the dynamo effect.

Dynamo effect theory

Figure II.2 illustrates the dynamo effect:

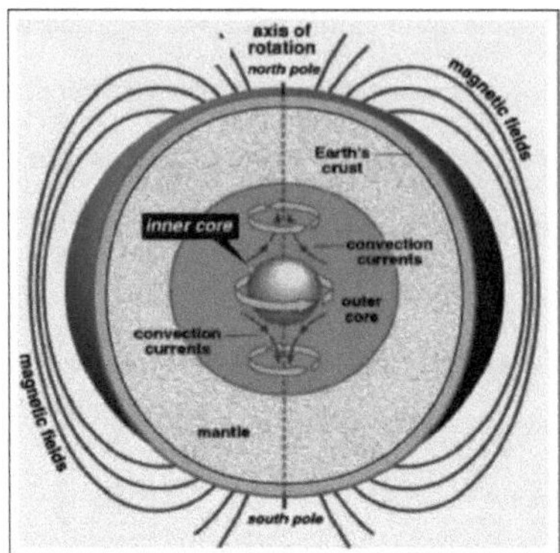

FIGURE II.2 - *Diagram of the Earth's inner structure, showing the inner and outer cores and the convection movements (arrows) that give rise to the magnetic field. Science. Cycles*

The dynamo effect suggests that the Earth's magnetic field is created and maintained by an induction process. Intense electric currents flow in the outer core, which has very high electrical conductivity (outer core: the liquid part of the core located between $r = 1300$ and 3500 km.

It is now assumed that the core is a combination of iron (Fe) and nickel (Ni), both good electrical conductors. Even if the core were made up of less conductive elements, the enormous pressure regained could press the electrons together to form free-electron gases of satisfactory conductivity.

The magnetic source is illustrated by the self-excited model. That is, a fluid of high conductivity moves in a complex motion and electric currents are caused by chemical variations producing a magnetic field (From T.Rikitake, Electromagnetism and the Earth's Interior", Elsevier North-Holland, 1966).

II.1.6. b The Transitional Field

It can be described as an additional field (transient field), generated by currents in the upper atmosphere and magnetosphere. This component undergoes two types of variation, which we summarize briefly as follows:

Diurnal variations

They are characterized by a low amplitude of around 30 to 40 Gamma, but have a periodicity of the order of a day; they reach a maximum around "noon", local solar time (the amplitude of the diurnal variation remains greater in summer than in winter).

Rapid or transient variations

Figure II.3 shows rapid and transient variations.

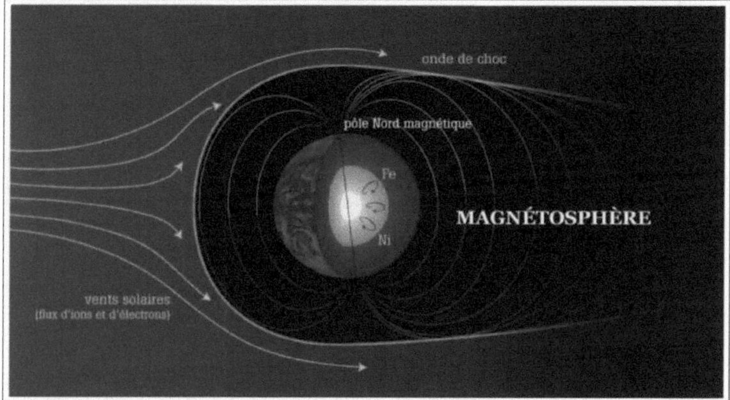

FIGURE II.3 - *Transient and internal field illustration, (www.laurinemoreau.com).*

rapid or transient variations are short-lived, generated by solar activity, with low amplitudes, but in the case of magnetic storms, they can reach up to 2000 gamma. The causes of the transient field are external to the globe, as it is clear that their periods are those of astronomical or astrophysical phenomena: rotation of the earth and sun, revolution of the earth, activity cycle of the sun.

The sources of the daily solar variation are thought to be electric currents circulating in the illuminated part of the ionosphere: the gas in the rarefied layer, ionized by solar radiation, is animated by horizontal movements under the effect of tidal forces of essentially thermal origin (and possibly gravitational: Moon-Sun influence): the movement of the conducting gas in the Earth's magnetic field induces electric currents, hence a magnetic field, which varies with the number of ions and their speed. Other variations can be attributed to various phenomena: variations in the solar wind and interplanetary field, electric currents (equatorial current ring).

11.1.6. c Anomalous field

The induced source, due to the induced magnetization of the rocks, is of weaker intensity and represents the anomalous magnetic field; it is superimposed on the main magnetic field. The aim of magnetic prospecting is to identify anomalies in the anomalous field.

II.1.7   Rock magnetism

Most of the elements that make up rocks are weakly or highly magnetic. Magnetic susceptibility k is the essential physical parameter. The response of rocks and minerals is conditioned by the quantity of magnetic minerals present; the latter have much higher k values than other minerals (magnetite, ilmenite, pyrrhotite). Significant differences are thus apparent in the main field, as a result of variations in the magnetic mineral content of rocks near the surface.

11.1.7. a Remanent magnetization

The magnetization of a rock is said to be remanent when it persists in the absence of an external magnetic source. It results from five types of magnetization:
1. thermoremanent magnetization (TMA), an essential mechanism in rock magnetization;
2. isothermal remanent magnetization (ART) ;
3. chemical magnetization (ARC);
4. detrital magnetization (ARD);
5. viscous magnetization.

The remanent mechanism of a rock can be important and have a polarity very different from the current field.

11.1.7. b Induced magnetization

Induction is much more important than remanence, except in a few rare cases (basalt, certain minerals). Magnetization intensity J is proportional to the applied field T :

$$J = kT$$

With k the magnetic susceptibility and T the applied field. The magnetic susceptibility of a rock generally increases with the percentage of magnetite and ilmenite it contains, and varies with the field T of ordinary temperature and with temperature for T a constant external field. The following table gives some average susceptibility values for minerals and rocks Magatte Fari [1995].

Minerals and rocks	Magnetic susceptibility (x 10^6)
Magnetite	>100.000
Ilmenite	30.000
Pyrrhotite	7.000
Hematite	150
Augite	150
Wolframite	210
Diorite	3.000
Peridotite-Dolerite	8000
Orthogneiss	1000 à 1500
Schist	100
Clay	200
Sandstone	0 à 150
Limestone	0 à 10
Anhydrite and gypsum	1 à 10

TABLE II.1 - *Susceptibility values for some minerals and rocks (after LASFARGUES (1996))*

We can see that sedimentary rocks and evaporites have the lowest average susceptibilities, while basic igneous rocks have the highest. Magnetic induction is the total field outside a body. It represents the resultant of the applied field and magnetization:

$$B - T + 4\pi.J$$

B is expressed in Tesla (SI unit) or in gauss (u.e.m or c.g.s system), the practical unit.

is gamma- γ-(lγ=10⁻⁵ gauss=10⁻⁹ Tesla).

The ratio of induction B to T, the field that causes it, defines the magnetic permeability μ. Equal to I in a vacuum, it is linked to the susceptibility k by the relation :
$$\mu - 1 + k$$

II.1.7. c Main types of magnetism

Materials (mineral elements, rocks) can be classified into three groups according to their magnetic properties: diamagnetic, paramagnetic and ferromagnetic. When the magnetic susceptibility of a material is negative (k<0), it is said to be **diamagnetic**; the intensity of induced magnetization opposes the inducing field. This property is weaker than other forms of magnetism. This category includes most gases, water, oxides, many metals (gold, mercury, silver, copper, lead) and minerals such as diamond, graphite, quartz, calcite, galena, and almost all organic compounds.

Paramagnetic bodies have a positive magnetic susceptibility (magnetic K>0). They lose their magnetization as soon as the external field disappears, as is the case with most rocks. Examples include limonite, hematite, pyrite, dolomite and spodumene.

Ferromagnetic bodies are the least numerous, but have the highest magnetic susceptibilities. Iron, cobalt and nickel make up this class. Chromium and manganese, paramagnetic in the free state, form ferromagnetic combinations with numerous metalloids: *MnBi, CrO, CrTe*.

Figure II.4 shows a ternary diagram of ferromagnetic minerals:

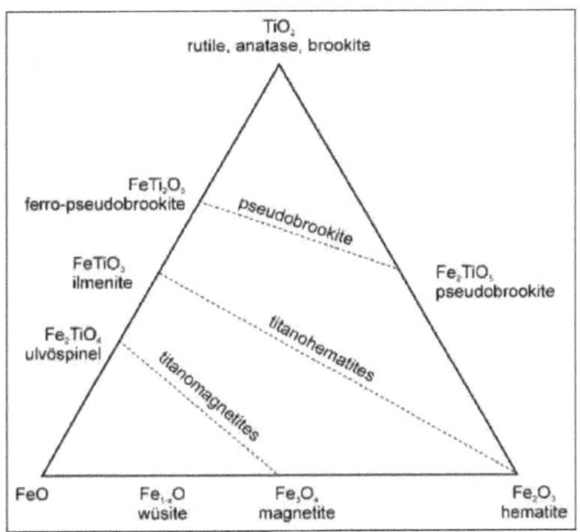

FIGURE II.4 - *Ternary diagram of FeO - Fe O -TiO₂₃₂ ferromagnetic minerals*

The minerals responsible for the magnetic properties of rocks mainly belong to the ternary system whose vertices a, b, c each correspond to *(a : FeO; b : Fe2O3 ; c TiO).2*

Other minerals include pyrrhotite and iron oxy-hydroxides (goethite *FeOOTa* and lepidocrosite *FeOOTγ*).

The Curie point is the temperature at which the magnetization of a ferromagnetic substance becomes negligible. For magnetite *Fe O₃₄* for example, the Curie point is 578 C, its density is 5.20 and its saturation magnetization at ordinary temperature is *480uem/cm³* .

II.1.8   Magnetic prospecting

Magnetic prospecting is based on the study of magnetic anomalies associated with rocks located in the crust Jacques Dubois [2011]. It involves surveying the total intensity of the Earth's magnetic field along certain profiles spread over a given surface or area. It can be carried out on land, at sea or in the air (airborne) H.Shout [2004].

II.1.7. a Fields of application

The fields of application of the magnetic method are :

1. **Geology:**

   ⇒ The reconstruction of past tectonic plate movements on geological time scales (paleomagnetism) and the periodic inversion of the earth's geomagnetic field in volcanic and sedimentary rock sequences (magnetostratigraphy);

   ⇒ As a support for geological mapping in areas where rock outcrops are rare, the magnetic map can be used to deduce the geological environment. Detecting faults and fractures (geological mapping)

2. **Mining exploration :**

   ⇒ Direct detection of metal deposits (magnetic iron, magnetite) and asbestos deposits (fibres are closely associated with magnetite and are found in very basic rocks);

   ⇒ The direct detection of Nickel associated with basic rocks, and mineralization generally associated with structures (faults, folds, intrusives, etc.) ;

   ⇒ Kimberlites (containing 5 to 10 percent iron oxides);

   ⇒ Interpolate between outcrops without drilling or digging.

3. **Petroleum exploration:** the study of sedimentary basins based on anomalies caused by basement structures or topography, and the indirect detection of structural traps (faults, folds).

4. **In hydrogeology:** groundwater research (fracture location and aquifer geometry).

5. **In archaeology:** the detection of buried metallic bodies at last highlights magnetic anomalies that can testify to vestiges over the last millennia.

6. **Environmental:** to map contaminants or polluted sites;

7. **Civil engineering:** to detect buried objects containing large quantities of iron. For example, a civil engineer might want to check that there are no wrecks of metal barges at the bottom of a river, or look for unexploded bombs buried in the

sediment before the construction of a port.

II.1.8. b Advantages, disadvantages and limitations

1. **Benefits**

   ⇒ inexpensive, light and fast method (gradiometry);

   ⇒ measuring the vertical gradient of the magnetic field eliminates the need for complex processing and the need to take into account temporal variations in the field;

   ⇒ total field measurement provides a more realistic representation of the magnetic field.

2. **Limits**

   ⇒ oxidation of ferromagnetic objects can sometimes reduce the possibility of detecting them, as they lose their magnetic character ;

   ⇒ solutions are not unique, and it's hard to tell the difference between all the objects that can be found on a polluted site (shells, drums, carcasses, refrigerators, manhole covers...) from their magnetic signatures alone.

3. **Disadvantage**

   ⇒ measuring the magnetic gradient is less accurate than measuring the total magnetic field;

   ⇒ total magnetic field measurement is more expensive, as it requires more complex processing before interpretation (but is richer in information).

   ⇒ the magnetometer is highly sensitive to the environment (power lines, fences, railroads, vehicles, magnetically saturated terrain) and to temporal variations in the natural magnetic field (in total field) Golle Olivia [2017].

Objects	Diastance (m)	Anomaly (nT)

Metal fencing	3	16
Metal fencing	8	1 to 2 (drowned in noise)
Railways	150	5 à 200
Railways	300	1 à 50

TABLE II.2 - *Maximum amplitudes of typical anomalies created by interfering objects as a function of their distance from the measuring instrument with very low magnetic noise and a sufficiently accurate magnetometer. Adapted from Breiner, 1999.*
These values can vary by a factor of 10 depending on object size, composition, orientation, magnetometer position and ambient field.

II.2   Methodology

Magnetometry, as a geophysical method, is designed to shed light on certain geological problems: locating and identifying certain rocks or geological structures hidden beneath the earth's surface, thus guiding geologists in their investigations. However, to overcome the problems associated with geology, the use of certain devices such as Magnetometers enables data to be acquired, which then undergoes appropriate correction and processing to give a geological meaning to the anomalies observed.

II.2.1   Devices used

The instruments used to complete this geophysical campaign are a Geometrics G-857 and a magnetometer proton-600, both of which are proton magnetometers.

These magnetometers are based on the proton precession measurement technique. This technique uses an induction coil to create a strong magnetic field around a hydrogen-rich fluid such as kerosene. This causes the rotational axis of the hydrogen protons to align or polarize with the newly applied magnetic field. When the current producing the polarizing field is interrupted, the protons begin to align themselves with the Earth's magnetic field; but, in doing so, it will momentarily precede the Earth's field at a specific frequency that is proportional to the intensity of the ambient magnetic field. This precession generates a small magnetic field which induces an alternating voltage in the induction coil previously used to generate the polarizing field. The relationship between the precession frequency of the induced voltage and the intensity of the earth's magnetic field is called the proton gyromagnetic ratio and is equal to 0.042576 *Hertz* per *nanotesla (Hz/nT)* (Geometrics manual operation). The proton precession magnetometer is one of the leading instruments for magnetic surveys, combining high accuracy with ease of use.

II.2.1. a G-857 Geometries

Figure II.5 shows a Geometrics G-857 proton precession magnetometer:

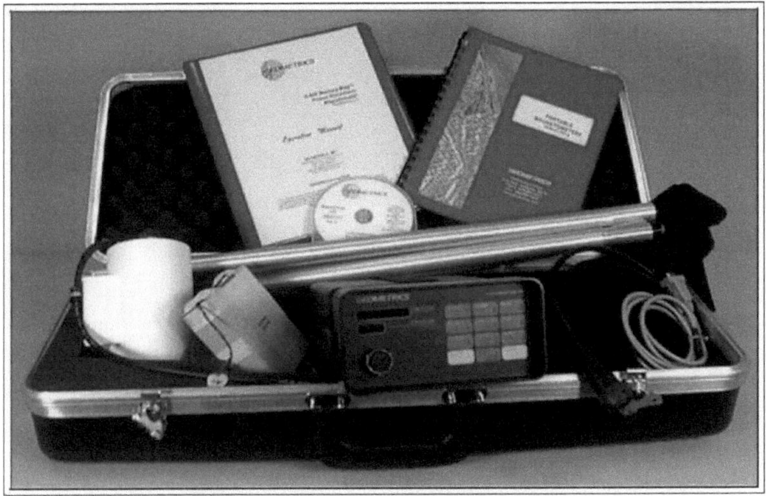

FIGURE II.5 - *G-857 Geometries magnetometer, comprising a console, 2 sensors and a rod.*

Depending on its particular configuration, the G-857 can be a man-portable magnetometer, a gradiometer measuring the field with two sensors, or a "base station" magnetometer. The G-857 can record the optional Garmin GPS in any configuration. As a hand-held instrument, it features a simple push-button operation and a built-in digital memory that stores 65,000 single-sensor measurements or 32,500 gradiometer readings. This eliminates the need for the user to physically write down data in the field, eliminates transcription errors and, most importantly, enables the use of computers to automatically record and process magnetic survey data.

The G-857 can also record data automatically at regular intervals, so it can be left unattended to monitor diurnal changes in the earth's magnetic field.

The G-857's measurement data can also be entered directly into an external computer. The time of day generated by the magnetometer's internal clock is recorded with each reading taken in either mode.

A single connector is used for sensor signal input and data output. Physically, the G-857 is compact and lightweight. It is weatherproof and operates over a wide temperature range. It is powered by either an internal rechargeable lead-acid gel-cell battery or an external 12-volt battery. Unlike other proton precession magnetometers, the G-857 has an internal programming switch that allows you to modify the magnetometer's cycle times to ensure that it will operate correctly anywhere in the world.

II.2.1. b Magnetometer proton-600

Figure II.6 shows a Magnetometer proton-600 proton precession magnetometer:

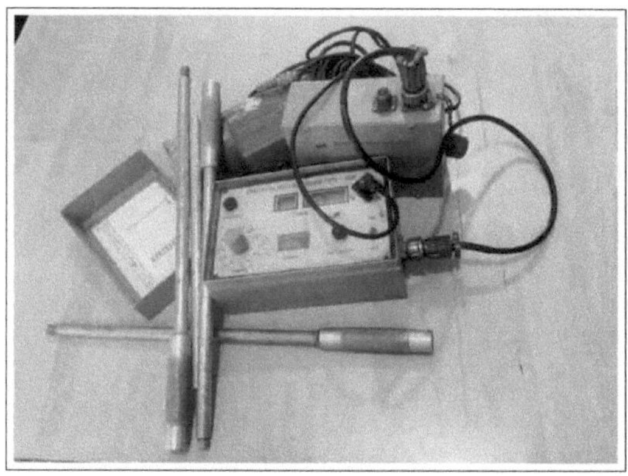

FIGURE II.6 - *Magnetometer proton-600, comprising console, rod, external battery and sensor*

This magnetometer operates in manual mode, which means that the operator must have a notebook in which to record the measurements, and a clock to mark the time at which the measurement is taken, as well as noting the date on which the measurements were taken. However, this magnetometer is equipped with a *Tuning* button to set the signal within the range of known values for a given zone. Here in Katanga, for example, the range is 31,000 *nT* to 32,000 *nT*. The *Signal* table lets you check signal quality and battery level, this time by clicking on the *Batt.check* button, and finally, the *Measure* button for taking measurements.

II.2.1. c GPS (Global Positioning System)

A geolocation tool is undoubtedly of paramount importance when conducting a geophysical campaign, and Garmin GPS units were used during this geophysical survey. They are illustrated in figure II.7 below:

FIGURE II.7 - *The Garmin 64χ and* 64

It takes the geographic coordinates of each reading point in the field and stores them in its internal memory. We used two Garmin 64 and *64 X* GPS units, with an accuracy of three meters.

II.2.2   Preparing the magnetic survey

This preparatory phase is essential to the success of the survey. It requires a clear idea of the geological targets to be reached and the types of anomalies that may be associated with them. Examination of aerial photographs and local geology, as well as knowledge of the inclination of the earth's field, provide a rough idea of the shape and size of the anomalies associated with the desired structures, and thus help plan the survey. To ensure that the signatures of the anomalies are as legible as possible, the profiles should ideally be perpendicular to the elongation of the presumed structures, intrusions and fractures, and parallel to the NS Magnetic Meridian to provide more information on the anomalies. It is essential that each anomaly is highlighted by several measurements; to do this, we tighten the measurements on each profile. Periodic returns to an arbitrarily chosen, but magnetically quiet, base can eliminate diurnal variation if the precision required for the survey demands it. The technique of magnetic prospecting consists in searching for and locating rocks, formations and deposits by the magnetic anomalies or local variations they produce in the earth's magnetic field. This method is widely used in mining exploration, and involves detecting mineralized zones generally associated with structures (faults, folds, intrusions) Olivier [2005].

II.2.2 .a Device (network of lines or profiles)

Completing a campaign also depends on the establishment of a line network device, which will serve as a sampling model for the entire survey. Figure II.8 below shows the device used for the survey:

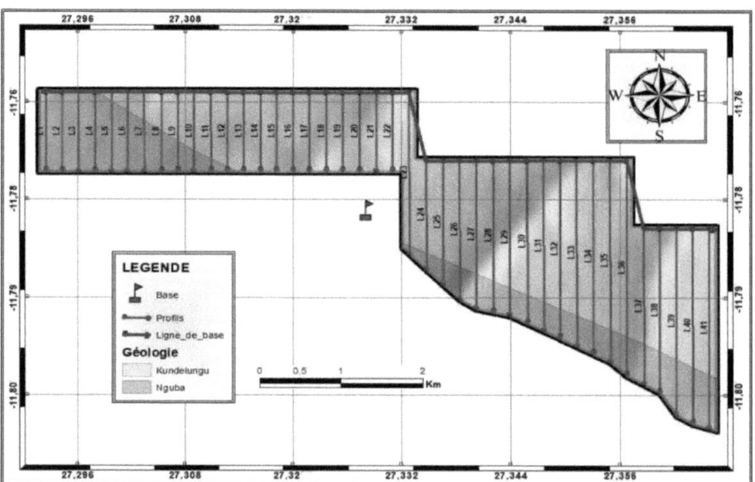

FIGURE II.8 - *Map of the network of lines surveyed during the data acquisition phase.*

According to a mathematical description, our study area takes the form of an irregular polygon with a surface area of 13.795009 km^2.

The pre-established profiles for magnetic surveying follow a NS orientation with an equidistance of 200 m from each other according to longitude, making a total of 41 profiles to be surveyed. The stations vary according to latitude, with an equidistance of 10 m, and the profiles are connected to an EW baseline made up of the set of points located at the northern extremity of each line or profile. These profiles vary in length from 1 km to 3 km.

II.2.2. b Measurement and data acquisition

Measurement and data acquisition must comply with a number of standards set by geophysicists for ease of use during the data processing and interpretation phase.

II.2.2. b.1 Execution of measures

Depending on the network of lines or profiles defined above, it is preferable for field work and subsequent interpretation to take constant-pitch measurements at stations located on profiles or straight lines more or less parallel to each other and perpendicular to the geological formations, as computer processing and interpretation will be easier and quicker.

II.2.2. b.2 Data acquisition

During the magnetic survey, the two proton magnetometers G-857 Geometrics and Magnetometer proton-600 provide an absolute measurement in nanotesla (nT) of the total magnetic field strength F at a point.

At the base

The location of a fixed base for a geophysical survey is of great importance, as the data obtained from it will help to eliminate certain effects linked to drift and temporal variations. With this in mind, figure II.9 shows the Magnetometer proton 600 type magnetometer with a fixed base and a fixed operator.

FIGURE II.9 - *Basefix with proton-600 magnetometer and fixed operator.*

Since the choice of field base location is very important for magnetic prospecting. The base station was set up at a point with UTM coordinates *(X:* 535759 and *Y:* 8697705) next to the area to be covered, in a magnetically quiet environment with a low magnetic gradient, outside any anomaly zones, far from any scrap metal and high-voltage power lines. An environment is said to be magnetically quiet if the magnetic field variation does not exceed 50 nT per hour Chouteau [2002], we point out that our readings at the base did not exceed 40nT per day. In order to evaluate the drift and get rid of the field of diurnal variations in the magnetic field, we used a Magnetometer proton 600 base station in manual mode, with a fixed operator at the base who took measurements manually, taking one reading every 10 or 15 seconds.

Measurements at the base station began 30 to 45 minutes before the mobile operators started measuring in the field. The base station operator stopped data acquisition after the mobile operators. These measurements at the base station serve as a reference for those taken during magnetic prospecting in the field.

In the field

In the field, mobile operators take measurements of magnetic field intensity.

Figure II.10 shows a field survey:

FIGURE II.10 - *Field survey by two mobile operators.*

Our lines or profiles have been programmed to intersect geological formations, and both ends of a profile are marked with a GPS (Garmin 64 and 64S). A profile is marked by a sticker post placed at each end (on a tree) and bearing the name of the profile, the alignment of the profile is obtained by GPS sighting and the device map imported into the GPS. As it was impractical to return to the same base station on a regular basis due to the great distances covered during the survey, the baseline will enable the measurements of each line or profile to be corrected.

II.2.3   Data correction and processing

II.2.3. a Correction of diurnal variation :

in prospecting, the first and last measurements of the day must be taken on the same base. If the total magnetic field is measured at the same point several times during the day at a given time interval, or on different days, deviations (drift) are recorded. As

47

magnetometry is concerned with spatial, not temporal variations in the total magnetic field, it is necessary to eliminate the effect of the latter by means of appropriate corrections.

There are methods for correcting these variations. These are the loop method, the base station network method and the fixed base station method. All these methods assume that diurnal variations are roughly linear over short periods of time, and that they are identical within a relatively small area.

We will therefore correct the drift profile by profile, using the following formula:

$$Vcor - Vlu \pm TD * (T - T_1)$$

with :

$\Rightarrow$ Vcor: corrected loop value in nanoteslas (nT),

$\Rightarrow$ Vlu: value read from nanoteslas loop (nT),

$\Rightarrow$ T_1 : reading time for first time to base in seconds (sec),

$\Rightarrow$ T: Vlu reading time in seconds (sec),

$\Rightarrow$ TD: drift rate in nanoteslas per second (nT/sec); defined by :

$$TD = \frac{(V_2 - V_1)}{(T_2 - T_1)}$$

with :

$\Rightarrow$ V_2 and V_1 respectively measure the start and end at the base station of a given loop and T_2 and T_1 their corresponding reading times.

II.2.3. b Correction of the difference between the two magnetometers :

although both magnetometers measure the total magnetic field intensity, there is a discrepancy between the two readings at a given time at the base. To correct this discrepancy, we will calculate only for the baseline *(BL)* the reading corrected for temporal and instrumental variations *(Lcor$_i$)*, which is equivalent to the sum of the levelled reading *(Lniv$_i$)* and the temporal correction *(Ct$_i$)* at any given baseline to obtain the actual values *(Lcor$_i$)* Michel Allard [1999] ;

$$(Lcor_i) = (Lniv_i) + (Ct_i)$$

This correction *(Ct$_i$)* eliminates temporal variations in the magnetic field. It corresponds to the difference between the IGRF reference total magnetic field reading at the base *(Lref)* and the magnetometer reading (fixed base) at any time *(LBF$_i$)*;

$$(Ct_i) = (Lref) - (LBF_i)$$

The levelled reading *(Lniv$_i$)* corresponds to the sum of the correction of the difference between two *CDM* magnetometers$_i$ and the reading of the moving magnetometer at any base *(LMMi)*;

$$(Lniv_i) = C_{DMi} + L_{MMi}$$

The correction of the difference between two magnetometers C_{DM} is directly proportional to the product of ΔEc and ΔT_{n-l} , and inversely proportional to the sum of ΔTf_l and Ec_l .

⇒ ΔEc: the difference between the last two and first readings of fixed and moving magnetometers at the same instant ;

⇒ ΔT_{n-l} : the difference between the times at a given base and the respective loop start times;

⇒ ΔTf_l : the difference between the start and end times of the loop under consideration ;

⇒ Ec_l : the difference between the first readings of the fixed and moving magnetometers.

II.2.3.c Loop correction with levelling :

Following correction of the difference between two magnetometers and of the profile-by-profile diurnal variation, if it turns out that the total magnetic field values are not identical. The baseline corrected values and the profile-by-profile corrected values at the baseline form loops (actual and observed corrected values).

So we'll have 41 loops, and the formula below will bring all the values measured on each segment back to the same level as the values initially measured on the baseline (there is only the spatial and not the temporal variation after the two corrections).

$$Vniv_i - Vlu_i - \frac{(M_i - V_i)}{Nb_t} * (Nb_i - Nb_0)$$

with :

> $\Rightarrow$ $Vniv_i$: the levelled value in any loop n_1 ;

> $\Rightarrow$ Vlu_i : the value read in any loop n_i ;

> $\Rightarrow$ V_i : the new value observed at a base station on the baseline (i) after diurnal correction profile by profile;

> $\Rightarrow$ M_i : the actual value at a base station (i) on the baseline after correction for the difference between two baseline magnetometers;

> $\Rightarrow$ Nb_t : the total number of stations in any loop n_i ;

> $\Rightarrow$ Nb_i : the number of the station taken in any loop n_i ;

> $\Rightarrow$ and Nb_0 : the number of the first station in any loop n).,

II.2.4   Data processing

11.2.4.   a Background noise filtering

Background noise filtering is used to get rid of erratic values due to the presence of steel objects in the ground or subsoil, and also geological noise due to the presence of heterogeneities within a single geological unit (no treatment exists for these cases). We used MagPick software to eliminate levelling errors and noise by manual filtering.

11.2.4.   b Pole reduction

Pole reduction consists in calculating the values that would have been obtained if the source of the anomaly had been located at one of the poles. Anomalies are then symmetrical and directly above the vertical bodies that generate them Michel Allard [1999]. This treatment has the advantage of :

⇒ Simplify the shape of anomalies by making them more symmetrical;

⇒ Position the maximum anomalies with respect to the abnormal source environment;

⇒ Enable better resolution of closely spaced anomalies.

The geological interpretation of the reduced field is therefore more intuitive than that of the measured magnetic field. Bipolar anomalies then become unipolar anomalies, centered on the structures that cause them. The use of pole reduction does, however, have its limitations:

⇒ the existence of remanent magnetism having a direction and sense different from

those of the current field is generally not taken into account by reduction to the off-center pole with respect to the source and not symmetrical. If the remanent field is collinear, in the opposite direction and stronger than the current field (inverse magnetization), the anomaly will be localized, but with the opposite sign;

⇒ chemical alteration processes that gradually reduce the magnetic mineral content

of rocks: the transformation of magnetite into pyrite or the destruction of magnetite due to hydrothermal alteration, observed on various geothermal fields in volcanic environments, are manifested by negative or weak anomalies Fabriol [2004].

II.2.3. c Enhancement of superficial anomalies

Among the techniques used to enhance anomalies, the first vertical derivative (or vertical gradient) and the second vertical derivative have proved very useful over the years Michel Allard [1999].

1. the first vertical derivative is used to refine the magnetic characteristics and

improve shallow structures over deeper ones. If after this first derivative the dominance of deep structures is still significant, it is important to consider a second vertical derivative.

2. the horizontal derivative highlights structures with a certain strike by accentuating gradients. It is normally calculated perpendicular to the strike or main structural direction.

Vertical and horizontal derivatives are effective treatments for delineating the crustal lineament structure and identifying the boundaries of different facies. They convert magnetic anomaly gradient bands into zero and maximum values respectively for easy identification and precise positioning (Philips, 2000 ;Grauch, 2001 ;Verduzco et al., 2004 ;Wang et al., 2021).

We will only use the horizontal gradient, as it is adopted synthetically to identify crustal lineament structures and limit the lateral extension of different units. To better understand possible fault lines and large-scale structures, magnetic anomalies are continued upwards at 5 and 10 km, respectively (Blakeley, 1996). The horizontal gradient is widely used to locate subsurface discontinuities from both magnetometric and gravity data (Fedi and Florio, 2001). The maxima of the horizontal gradient coincide with the positions of these discontinuities Michel Allard [1999]. the SingProc software enabled us to perform this treatment.

II.2.5   Correction procedure and data processing

After any magnetic prospecting campaign, a certain amount of processing must be applied to the data, depending on the acquisition method chosen, before we can move on to the interpretation phase. For this reason, we applied the line array method with a baseline. According to the IGRF, during the survey period the average values at the base were :

$\Rightarrow$ 31352.5 nT for the total magnetic field;

$\Rightarrow$ -2.7° for declination;

$\Rightarrow$ And -47.3° for inclination.

These values were used in further corrections and processing to correct the difference between two magnetometers, filtering and pole reduction. The data thus collected underwent the following corrections: The diurnal variation correction was applied to the data using the MagMap software and the data from the fixed base; After the diurnal variation correction, we then applied a correction to the difference between the two magnetometers with levelling, to bring the two magnetometers (Geometrics G-587 and Magnetometer Proton-600) back to the same level. This correction only concerned the base line, which was then used as the basis for the loop correction with levelling on each line or raised profile. Following these corrections, we used :

1. SignProc software to perform background noise filtering to get rid of erratic values, and also apply RTP to the filtered data;

2. RTP data were imported into MagPick software to estimate location, relative roof depth and body tilt direction;

3. Finally, the horizontal gradient was applied to the pole-reduced data, again using MagPick software.

Our interpretation is therefore based on these results for total magnetic field, pole reduction (RTP) and horizontal gradient.

II.2.6   Geophysical data interpretation techniques

The geophysicist looks for a model to account for the anomaly. The shape of the anomaly curve, profile or iso-anomal, gives us an idea of the disturbing body.
Qualitative interpretation of magnetometric data involves assessing the geometry and apparent magnetic susceptibility of the anomalous source, based on the characteristics of the anomalies observed on the magnetic profiles. These characteristics are: amplitude, wavelength and H.Shout shape [2004].
Quantitative interpretation was carried out using Geometrics' MagPick software to calculate the depth to roof, location, amplitude and direction of inclination of the body

buried in the subsurface.

II.2.7   Brief overview of the operating phases

Figure II.11 shows a summary of the operating phases that led to the realization of this project.

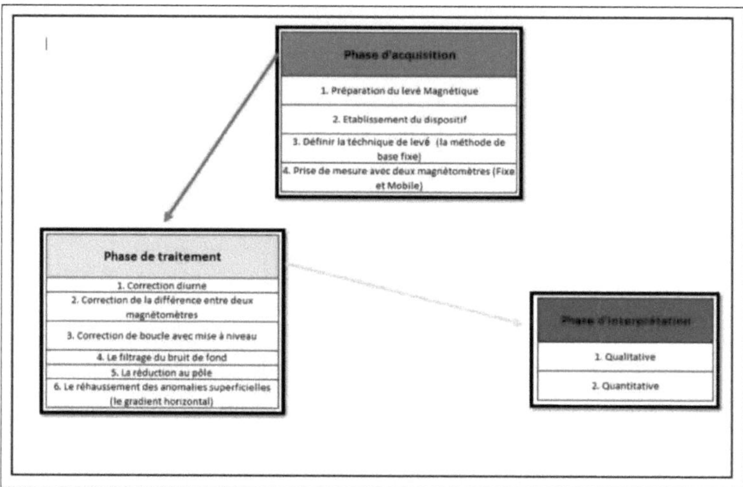

FIGURE II.11 - *The operating phases used in this project.*

Figure II.11 shows three important phases in this geophysical prospecting project: the acquisition phase in green, the processing phase in yellow and the interpretation phase in red.

Total magnetic field intensity values are presented and interpreted in two forms: in profiles and in plan Michel Allard [1999]. In this work, we have opted for the presentation of data in plan and profile form. The interpretation technique used defines or delimits zones with a certain number of common magnetic characteristics, so as to be able to define the different magnetic units, which in most cases correspond to the different lithologies found in the geological formations, and then detect boundaries and anomalous zones.

Chapter 3: PRESENTATION AND INTERPRETATION OF DATA

In applied geophysics, the interpretation of results involves two different aspects corresponding to two successive phases (Baranov, 1957). The first phase is a detailed analysis of the data obtained by each of the methods applied (Processing) and the second is the synthesis of geophysical and geological data.

In this chapter, we exploit these two phases using raw data, reduced to the pole, and the horizontal gradient.

III.1   Magnetometric mapping

As the first phase of interpretation involves a detailed analysis of the data obtained for each of the methods applied in the processing. In this section, we present the results of these treatments in turn.

III.1.1   Total magnetic field map

Since magnetic prospecting is concerned with the spatial variation of the magnetic field at each sampling point, the values obtained are sometimes affected by noise. The result of this operation is shown in figure III.1.

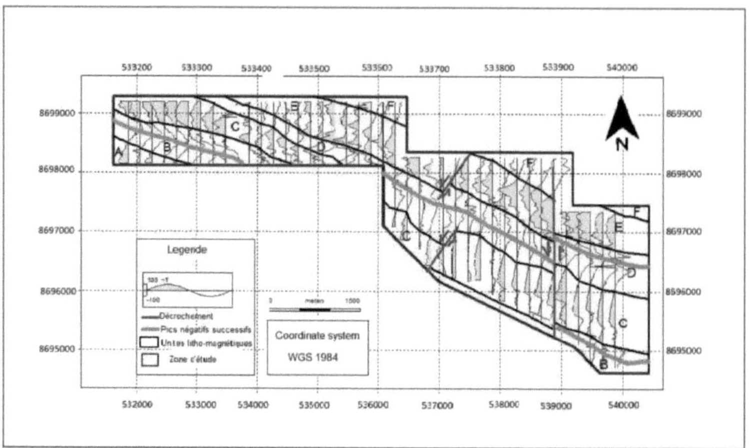

FIGURE III.1 - *The total magnetic field map showing the litho-magnetic units A, B, C, D, E, F characterized by either yellow (for positive amplitudes) or white (for negative amplitudes) colors and delimited by a black line, the orange banks correspond to successive negative peaks, the dark red lines mark a dip and the red arrows indicate the direction of movement of the litho-magnetic units.*

Figure III.1 shows magnetic field amplitudes ranging from -429.331 to 429.331 nT, with alternating positive and negative amplitudes from one lithomagnetic unit to another (A+, B-, C+, D-, E+, F-). Successive negative peaks are in units B and D, and the southeastern part is marked by three stalling structures, two of which are oriented SSW-NNE (stalling of units E, D and F, D, C) and the other almost N-S (stalling of units B, C, D, E).

III.1.2   Magnetic map of the reduced field at the pole

The change in orientation of the Earth's magnetic field affects the shape and amplitude of an anomaly. In order to compare anomalies detected at different locations on the globe, and thus simplify interpretation Michel Allard [1999], pole reduction is used. The result is shown in figure III.2.

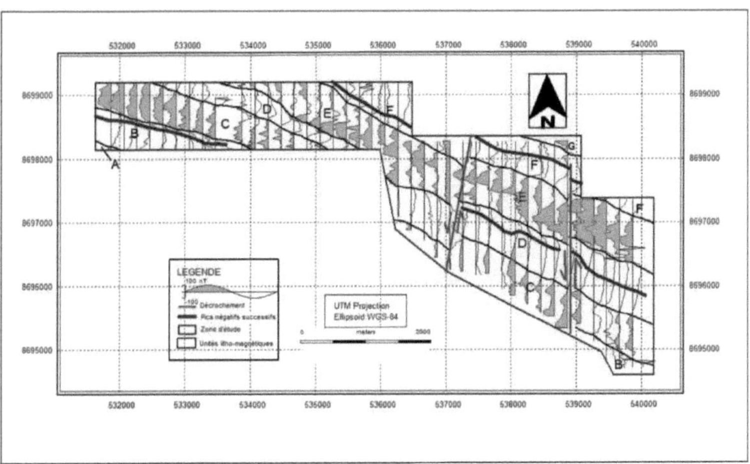

FIGURE III.2 - *The reduced-field map at the pole showing litho-magnetic units A, B, C, D, E, F characterized by either water-green (for positive amplitudes) or white (for negative amplitudes) colors and, delimited by a black line, dark-red banks correspond to successive negative peaks, purple lines mark a drop-off and purple arrows indicate the direction of movement of the litho-magnetic units.*

Based on amplitude variation, the result of pole reduction has enabled us to highlight seven lithomagnetic units: A, C, E and G have positive amplitudes, while B, D and F have negative amplitudes. Units B, D and F contain successive negative peaks and the south-eastern part of the zone is marked by two stalls, one of which is oriented SSW-NNE (stall affecting units E and D) and the other almost N-S (stall affecting units B, C, D, E and F). The amplitude after RTP is in the range -396.089 to 396.089 nT in figure III.2.

III.1.3   Horizontal gradient magnetic map

Strictly speaking, the horizontal gradient is the derivative of the X and Y components of the magnetic field. Its application to pole-reduced data enables us to delineate the different litho-magnetic facies by analyzing the amplitude variation. Figure III.3 shows the results obtained after this treatment.

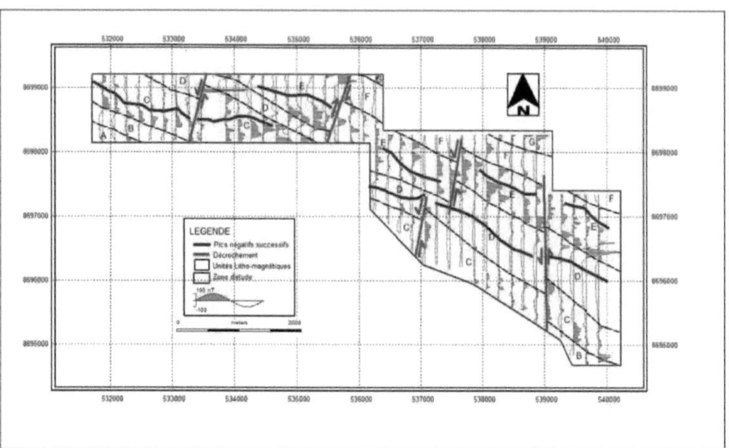

FIGURE III.3 - *Horizontal gradient map showing litho-magnetic units A, B, C, D, E, F, G characterized by either green (for positive amplitudes) or white (for negative amplitudes) colors and delineated by a black dashed line, dark red banks correspond to successive negative peaks, purple lines mark a drop-off and purple arrows indicate the direction of movement of the litho-magnetic units.*

In Figure III.3, the amplitude of the horizontal gradient varies from -24.119 nT to 24.119 nT. The alternation of negative and positive amplitudes, which marks the transition from one litho-magnetic unit to another, allows us to restrict the sample to seven units: A+, B-, C+, D-, E+, F- and G+. Units C, D and E are characterized by a succession of negative peaks. The area in general is marked by five descending structures, four of which are oriented SSW-NNE and the other almost N-S.

In summary, the total magnetic field map includes values ranging from 30896.031 to 31754.69 nT, with litho-magnetic units ranging from A to F, with alternating positive and negative values. In the south-eastern part of the study area, interpretation has enabled us to define stall structures, two of which are oriented SSW-NNE, offsetting units C, D, F, and another oriented N-S, affecting units B, C, D, E. After RTP, field values vary from 31032.615 to 31824.791 nT, and the aforementioned stall structures are restricted to two structures. Finally, application of the horizontal gradient to the pole-reduced data leads to the identification of two further SSW-NNE-trending stall structures in the N-W part.

These structures affect litho-magnetic units C, D, E.

III.1.4   Some profiles of the south-eastern part of the zone

In this subsection, estimates of depth to roof, direction of inclination and body location were made using MagPick software. Thus, the bodies presented in this sub-section have a relative depth to roof of at most thirty meters, with an inclination towards the North-East and/or South-West. The point of origin of each profile corresponds to the southernmost point and, based on this point, the distance from the origin (Dist-orig) is given for each body in the tables.

Figure III.4 shows profile 22.

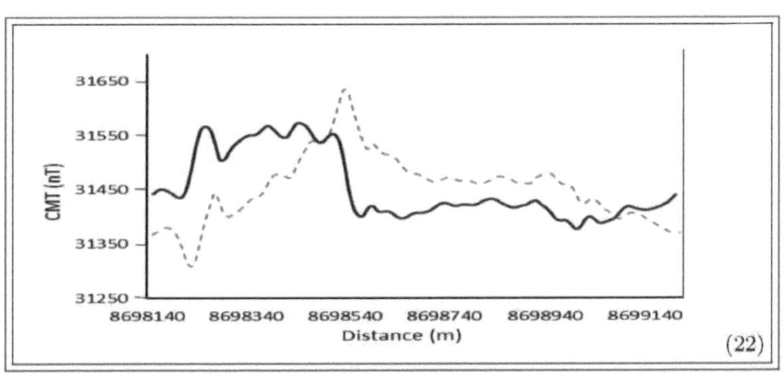

FIGURE III.4 - *The 22 profile of the dashed total magnetic field and that reduced to the continuous pole*

Considering the profile in figure III.4, the total magnetic field varies from 31300 to 31650 nT. Estimation of the location, relative roof depth, amplitude and direction of inclination on different parts showing abnormal variations provide the result shown in Table III.1.

TABLE III.1 - *Estimation results for profile 22.*

Profile	X	Y	Dist-orig	Depth to roof (m)	Amplitude (nT)	Tilt direction

| 22 | 536098,75 | 8698356 | 205 | 27,46 | -61,891 | NE |

Table III.1 shows the bodies inclined to the northeast with a negative amplitude of -61.891 nT and a roof depth of 27.46 m.

Figure III.5 shows profile 28.

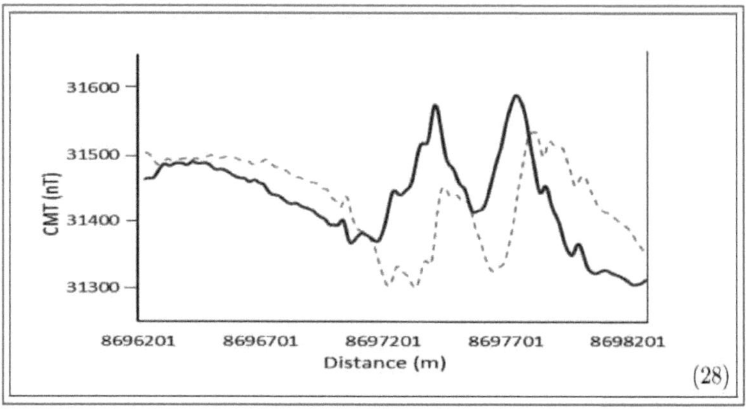

FIGURE III.5 - *Profile 28 (of filtered data in dashed line and that (of pole-reduced data in solid line.*

Figure III.5 shows the magnetic profile of filtered and reduced data with amplitudes in the range 31300 to 31600 nT. After estimation, the RTP profile provides results on location, relative roof depth, amplitude and body tilt direction as shown in Table III.2.

TABLE III.2 - *Estimation results for profile 28*

Profile	X	Y	Dist-orig	Depth to roof (m)	Amplitude (nT)	Tilt direction
28	537310,38	8697388	1167	28,72	-95,697	NE
28	537265,44	8697058	837	30,84	-34,666	NE

Table III.3 shows two bodies inclined to the NE with a mean amplitude value of 65.1815nT nT. The average depth of these bodies to the roof is 29.78 m.

Figure III.6 shows profile 29.

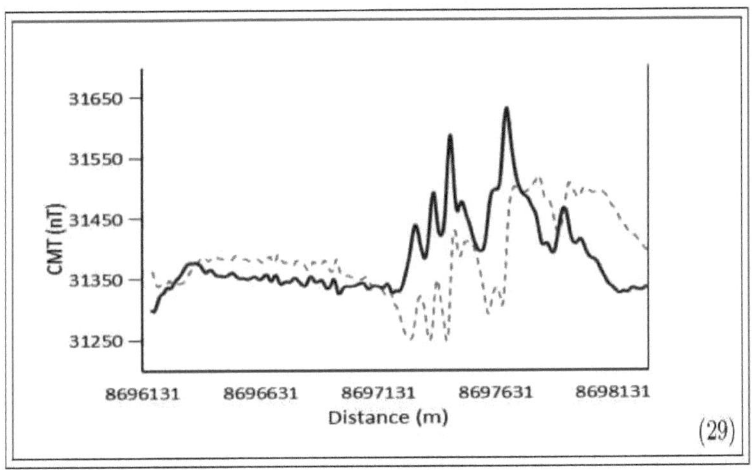

FIGURE III.6 - *Profile 29 (data filtered in dashed line and reduced to the pole in continuous line).*

Figure III.6 shows the magnetic profile with values in the range 31250 and 31650 nT. After estimation, the RTP profile provides information on the location, relative roof depth, amplitude and inclination direction of the body, as shown in Table III.3.

TABLE III.3 - *Estimation results for profile 29.*

Profile	X	Y	Dist-orig	Depth to roof (m)	Amplitude (nT)	Tilt direction
29	537438,13	8697438	1277	27,97	166,576	SO
29	537462,31	8696217	56	17,17	-59,143	NE

In Table III.3, we have identified two bodies using MagPick software, one inclined SW and the other NE. The average depth of these bodies at the roof is 22.57 m, with an average amplitude of 53.72 nT.

Basically, on the three profiles selected in the south-eastern part of the study area, we have five bodies resulting from the MagPick estimation. Of these, one is inclined to the south-west and four to the north-east.

III.2   Geological mapping using magnetometry

Magnetic mapping is often in line with geological mapping, as the quantity and distribution of magnetic minerals generally varies from one rock type to another. In this section, the aim is to establish a geological map based on the spatial variation of the magnetic field in the study area. We present in turn the maps obtained after filtering, pole reduction and application of the horizontal gradient, while matching litho-magnetic units to probable geological formations.

III.   2.1 Total magnetic field map draped over local geology

Figure III.7 below shows the total magnetic field map superimposed on the local geology.

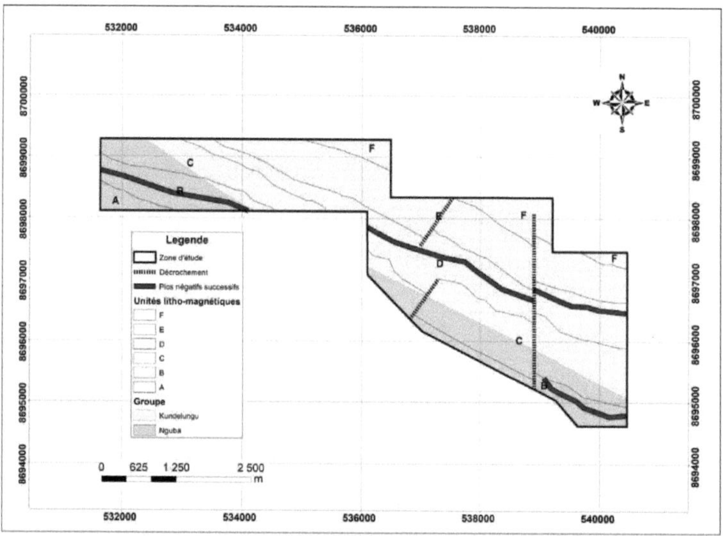

FIGURE III.7 - *The total magnetic field map draped over the local geology where light yellow corresponds to the Kundelungu subgroup, light orange to the Nguba subgroup and blue dashed lines correspond to litho-magnetic boundaries, black dashed lines to decays and dark red to successive negative peaks.*

In figure III.7, lithomagnetic units correspond to formerly undifferentiated geological formations and by correspondence, probably the Kipushi formation (Ng1.4) corresponds to lithomagnetic unit A, the Katete formation (Ng2.1) to lithomagnetic unit B, Monwezi (Ng2.2) to unit C, Lusele (Ku1.2) to unit D, Kanianga (Ku1.3) to unit E and the Mongwe (Ku2.1) and Kiubo (Ku2.2) formations remain undifferentiated to date. We note the presence of three senestial stalls in the south-eastern part of the study area, two of which are oriented SSW-NNE and one almost N-S. The successive negative peaks correspond to quartz veins observed in the field.

III.2.2   Map of local geology data reduced at the pole

The map of reduced pole data superimposed on the local geology is shown in figure III.8.

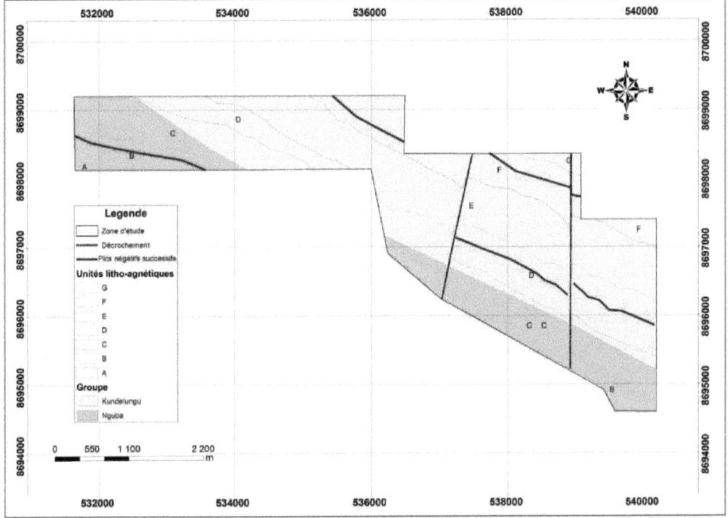

FIGURE III.8 - *The map of reduced magnetic data at the pole, draped over the local geology of the study area, where the dashed blue lines correspond to lithomagnetic boundaries, the black lines correspond to dislocations and the dark red lines to successive negative peaks. Light yellow corresponds to the Kundelungu subgroup, light orange to the Nguba subgroup.*

Figure III.8 shows an additional litho-magnetic unit, Unit G, which may correspond to the Kiubo formation (Ku2.1), and the two SSW-NNE-trending stalls restricted to a single senestial stall affecting the Monwezi (Ng2.2), Lusele (Ku1.2), Kanianga (Ku1.3) and Mongwe (Ku2.1) formations. The above-mentioned veins also occur at this level.

III.2.3   Horizontal gradient map draped over local geology

Figure III.9 shows the horizontal gradient map superimposed on the local geology of the study area.

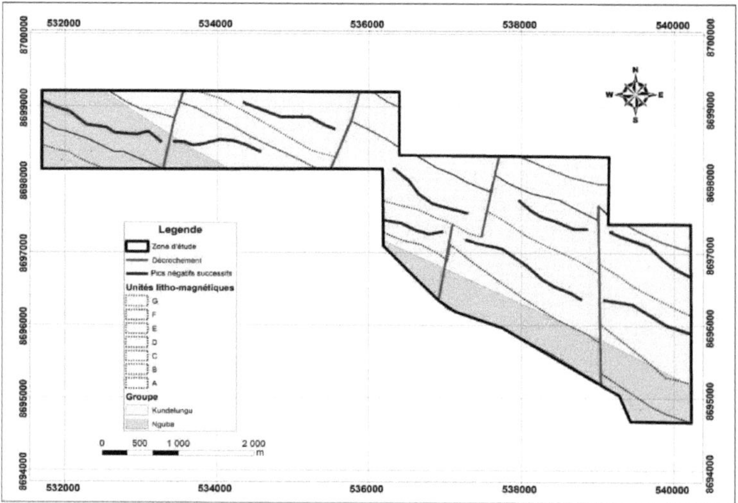

FIGURE III.9 - *Horizontal gradient map draped over the local geology of the study area, where dashed blue lines correspond to lithomagnetic boundaries, red lines correspond to fault lines and dark red to successive negative peaks. Light yellow corresponds to the Kundelungu subgroup, while light orange represents the Nguba subgroup.*

Figure III.9 shows an addition from a structural point of view. Applying the horizontal gradient to the pole-reduced data, instead of the three stall structures mentioned above, we have five stall structures affecting the zone. Of these, four are oriented SSW-NNE and one almost N-S. The veins highlighted in the previous subsections are also identified, but this time with additional veins, as we can see.

III.2.4 Correlation of litho-magnetic units with the litho-stratigraphy of the study area

Figure III.11 graphically presents the litho-magnetic units corresponding to the probable geological formations identified by magnetic approach in the study area. This correlation takes into account the formations predefined by François [1973] and Intiomale [1982].

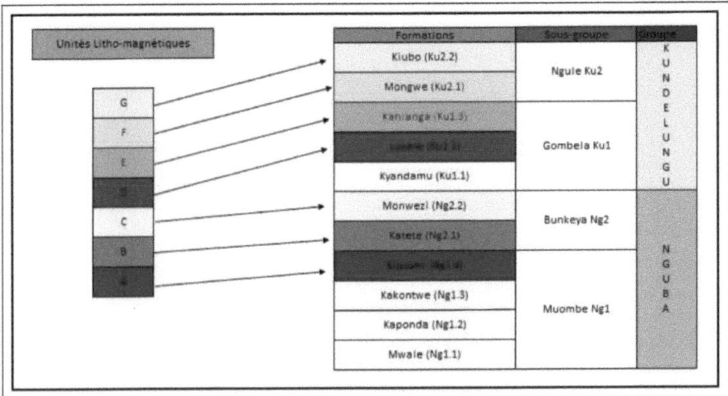

FIGURE III.10 - *Correlation of litho-magnetic units and geological formations.*

A simple and rapid compilation of the geological formations identified from the litho-magnetic facies led to the elaboration of figure III.10. Based on this correlation, we were unable to identify the Kyandamu Ku1.1 formation, as it was not mentioned in our study area by François [1973] and Intiomale [1982].

III.2.4. a Geological map obtained

Figure III.11 shows the proposed geological map after correlation of the litho-magnetic units with the litho-stratigraphy of the study area.

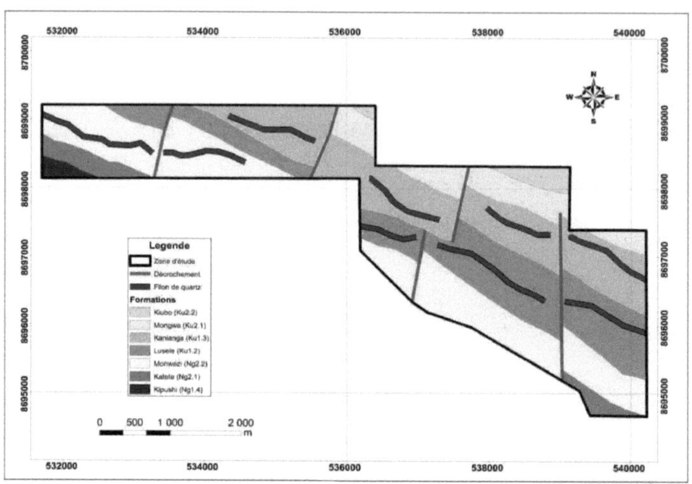

FIGURE III.11 - *The geological map obtained after correlation between litho-magnetic units and geological formations.*

This geological map shows 7 formations identified according to a given magnetic signal and existing bibliography. The north-western part is marked by two senestial decays oriented SSW-NNE and in the south-eastern part, three senestial decays this time two oriented SSW-NNE and one is oriented almost North-South. The quartz veins observed in the field could correspond to the negative peaks.

Examination of the magnetic data obtained after filtering, pole reduction and gradient led us to identify seven litho-magnetic units. These litho-magnetic units are subject to successive negative peaks and five stalls. We believe it is necessary to assign these litho-magnetic units to known geological formations in the area. By analogy with the undifferentiated local map, we are inclined to consider that the seven lithomagnetic units correspond successively to the Kipushi (Ng1.4), Katete (Ng2.1), Monwezi (Ng2.2), Lusele (Ku1.2), Ka- nianga (Ku1.3), Mongwe (Ku2.1) and Kiubo (Ku2.2) formations.

Negative peaks could be explained by the presence of quartz veins.

We emphasize that the Kyandamu Formation has not been differentiated, as mentioned by François [1973]. The Ng2.1, Ng2.2, Ku1.2, Ku1.3 and Ku2.1 formations are affected by senestial stall structures, four of which are oriented SSW-NNE and only one is oriented almost North-South.

III.2.5   Some geophysical profiles on geological sections

In this subsection, we present magnetic profiles superimposed on geological sections. These profiles are derived from the application of the horizontal gradient to the pole-reduced data. Superimposing these profiles on the geological sections enables us to check the delineation made in plan and locate the disturbing bodies on each characteristic profile.

Figure III.12 shows some geophysical profiles:

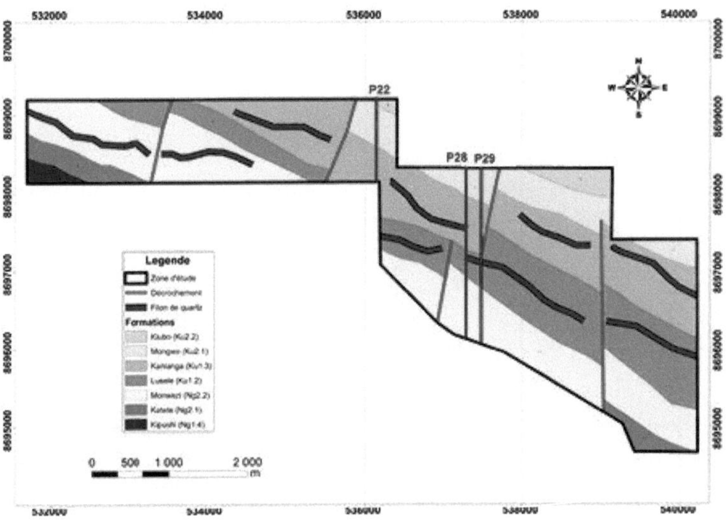

FIGURE III.12 - *Geological map obtained and some profiles of unstuck zones in blue.*

III.2.5. a Profile 22

Profile 22 superimposed on the geological section is shown in figureIII.13.

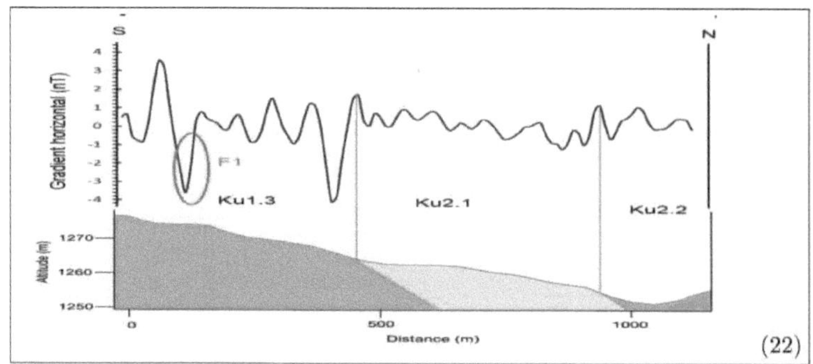

FIGURE III.13 - *Profile 22 of the horizontal gradient superimposed on the geological section, where the green ellipsoids represent buried bodies and the blue lines correspond to the boundaries of the layers.*

On figure III.13, we can see exactly as on the plan map the different formations identified and the gradient values vary between -4 nT and 4 nT and the body deduced corresponds to the quartz vein. .

III.2.5. b Profile 28

Figure III.14 shows profile 28

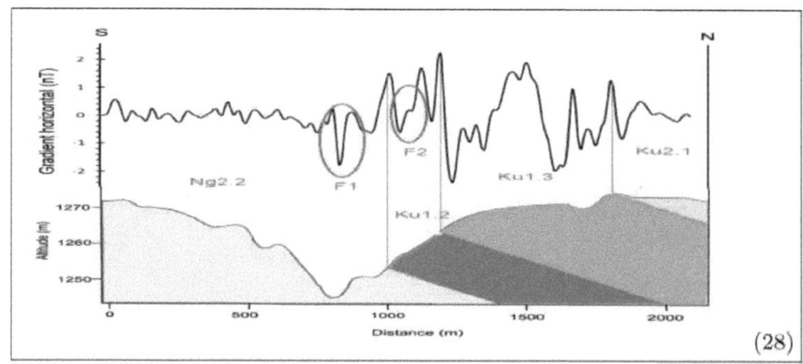

FIGURE III.14 - *Profile 28 of the horizontal gradient superimposed on the geological section. The green ellipsoid indicates the presence of a disturbing body, and the blue lines correspond to layer boundaries.*

The profile shown in figure III.14 is characterized by hori- zontal gradient values ranging from -2 nT to 2 nT. As mentioned in Table III.3, we have two bodies corresponding to vein quartz. Four formations are identified: Ng2.2, Ku1.2, Ku1.3 and Ku2.1.

III.2.5. c Profile 29

Figure III.15 below shows profile 29

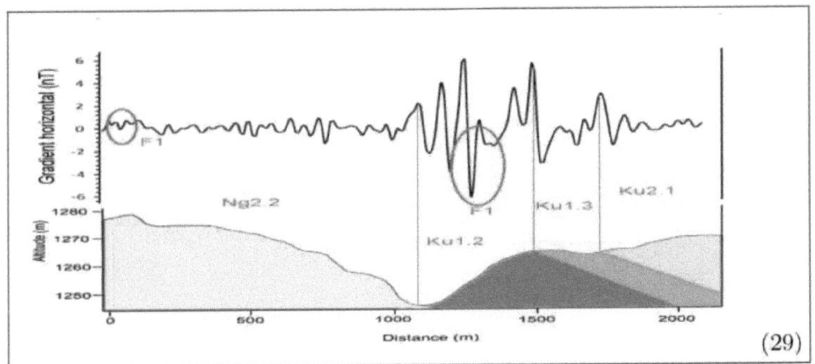

FIGURE 111.15 - *Profile 29 of the horizontal gradient superimposed on the geological section. The green ellipsoid indicates the presence of a disturbing body and the blue lines correspond to the layer boundaries.*

The profile shown in figure III.15 is characterized by horizontal gradient values ranging from -6 nT to 6 nT. As mentioned in Table III.3, we have here two bodies corresponding to vein quartz. The four formations identified are: Ng2.2, Ku1.2, Ku1.3 and Ku2.1.

This profile shows that each geological formation has its own magnetic character:

⇒ Monwezi (Ng2.2) is calm and very weakly magnetic, Lusele (Ku1.2) is restless and magnetically high, Kanianga (Ku1.3) is less restless and weakly magnetic, and Mongwe (Ku2.1) is calm and very weakly magnetic.

Roughly speaking, the profiles presented in this subsection are truncated and/or contain the above-mentioned vein bodies, which are deduced on the pole reduction curves from geometrics' MagPick software with a relative roof depth of up to 30 m. These profiles simply support the lithological contacts deduced in subsections III.2.1, III.2.2 and

69

III.2.3. In particular, we note that the Nguba Group formations are weakly magnetic compared with those of Kundelungu, with negative-amplitude surface bodies inclined to the NE. These bodies correspond to intra-formational quartz veins observed in the field.

GENERAL CONCLUSION

The results of the analysis of the magnetic field data after processing and correlation with the geology previously known in this study area led to the following observations:

Geologically speaking

⇒ Mapping using the magnetometric approach has enabled us to delineate (differentiate) the NW-SE trending formations of: Kipushi (Ng1.4); Katete (Ng2.1); Monwezi (Ng2.2); Lusele (Ku1.2); Kanianga (Ku1.3); Lubudi (Ku1.4) and Mongwe (Ku2.1), which belong to the Muombe (Ng1), Bunkeya (Ng2), Gombela (Ku1) and Ngule (Ku2) subgroups found in the Nguba and Kundelungu group.

⇒ Intra-formational quartz veins in the Ng2.2, Ku1.2 and Ku1.3 formations

⇒ The Ng2.1, Ng2.2, Ku1.2, Ku1.3 and Ku2.1 formations are distinguished from two others, including Ng1.4 and Ku2.2, by stall structures. Of these five senestial decays, 4 are SSW-NNE trending and only one is almost North-South trending. It should be noted that the Kyandamu formation (Ku1.1) could not be identified. However, it's important to stress that, at the stage we've reached with this work on the contribution of magnetometry to geological and structural mapping, it's difficult to say exactly what the rock is without calculating or measuring magnetic susceptibility and/or without upstream petrographic studies. And so, these observations can be supplemented and/or modified by detailed geological and structural mapping. Work on a comparative study of detailed mapping, structural and petrophysical mapping based on these results could provide more information and precision.

On the economic front

⇒ Our contribution is indirect, given that the results provided at this level have enabled us to highlight tectonic markers at the local level of dislocations that may be faults on a regional scale, in the knowledge that the latter would be the main vectors of mineralizing fluids. It is therefore important to have a clear idea on this subject before embarking on a subsequent phase of geochemical exploration or mining and drilling work.

Finally, magnetometry is of great importance and remains an essential tool for the production of a less evasive geological map that can provide the best possible information on the nature and characteristics of geological formations by integrating physical information.

Bibliography

AB Kampunzu, Jacques Cailteux, B. M. H. L. (2005). Geochemical characterization, provenance, source and depositional environment of sedimentary rocks of the argillaceous rocks (rat) and mines subgroups in the Neoproterozoic Katangese belt (congo): lithostratigraphic implications.

Bilolo (2016). Contribution à l'évaluation et à la gestion des impacts environnementaux des activités minières et industrielles anterieures dans le secteur minier de kipushi. *Exploration et Géologie Minière, University of Lubumbashi, 250.*

Brown (1979). Current status of the geochronology of Katanga. *Zaire.Ann.Soc/Géol. num :120,* (531-536).

Cahen (1970). Current state of the geochronology of Katanga. *Royal Museum for Central Africa,* 165(7-14).

Cahen, D. (1977). The conglomeratic complexes of the south-eastern edge of the Kibaran chain and their relations with the Katanguian strata of the Lufilian arc. *Royal Afrique central,* (111-135).

Cailteux (1981). La couverture katanguienne entre les socles de n'zilo et kapombo (rdc région de kolwezi). *Musée royal d'Afrique central, Série Num :8. Soc/Géol. num :87,* 48.

Chabu.M (1990). Metamophism of the kipushi carbonate hosted zinc-lead-copper deposit (shaba-zaïre). *Université de Lubumbashi,* 160.

Chouteau, M. (2002). Applied geophysics (magnetism). *École polytechnique de Montréal,* 98(7).

D. Delvaux, A. B. (2010). African stress model from formal inversion of focal mechanism data. 482.

Fabriol, H. (2004). Geophysical methods applied to geothermal exploration in volcanic island contexts (bibliographic synthesis). *Report BRGM/RP-53137- FR.*

François (1987a). L'extrême occidentale de l'arc cuprifère shabien. *(Etude géologique de Gécamines)Departement de géologie UNILU,* 65.

François, A. (1973). The western end of the Shabian copper arc. 120.

François, A. (1987b). Geological synthesis of the Shaba copper arc (Zaire).

Golle Olivia, C. P. (2017). Guide to geophysical methods for the detection of buried objects on polluted sites. *ADEME,* 124(24-25).

H.Shout (2004). Geophysics for geologists, vol.2. 65(6).

Intiomale (1982). Le gisement zinc-plomb-cuivre de kipushi (shaba-zaïre), étude géologique et métallogénique. *Dsc Thesis submitted to University Catholique de Louvain,* 170.

J. Batumike, WL Griffin, E. B. (2008). Lam-icpms u-pb dating of kim- berlitic perovskite: eocene-oligocene kimberlites from the kundelungu plateau, dr congo.

J. Cailteux, A. M. (2007). Base metal deposits hosted in Neoproterozoic sediments of western Gondwana.

J. Cailteux, ABH Kampunzu, J. B. (2005). Lithostratigraphic position and petrographic characteristics of the rat subgroup ("argillaceous rocks"), Neoproterozoic Katangese belt (Congo).

J. Cailteux, T. D. P. (2019). Complex mineralogical-geochemical sequences and alteration events in supergene ore from the cu-co luiswishi deposit (katanga, dr congo).

Jacques Dubois, M. D. e. J. P. C. (2011). *Geophysics (course and corrected exercises).* Dunod.

J.Batumike, M.C., K. A. (2007). Lithostratigraphy, basin development, base metal deposits, and regional correlations of the neoproterozoic nguba and kundelungu rock successions.

Kampunzu (1998). The mesoproterozoic kibaran belt system in africa: a key for the reconstruction of rodina supercontinent. *Gondwana Research 1,* (412-414).

Kanzundu (2020). The genesis of copper-cobalt and uranium mineralization at luisha principal (haut-katanga, drc): petrography, radiometry, geochemistry and metallogeny. *PhD thesis in geology, Unilu,* 241.

Kipata (2013). Brittle tectonics in the lufilian fold-and-thrust belt and its foreland. an insight into the stress field record in relation to moving plates (katanga,drc). *PhD Thesis,KU Leuven University,* 160.

Kokonyangi, Richard Armstrong, A. K. (2004). U-pb zircon geochronology and petrology of granitoids from mitwaba (katanga, congo): Implications for the evolution of the mesoproterozoic kibaran belt. 132.

Kpirgbene, W. (2016). Geological mapping by airborne magnetic method: application to two areas in the abitibi-temiscamingue region. 136.

Magatte Fari, K. N. (1995). Interpretation of geophysical data on the deep structure of the Senegalese sedimentary basin and the basement zone in eastern Senegal. *Université Cheihk Anta Diop de Dakar,* 159(78-81).

Michel Allard, D. B. (1999). *Geophysics applied to mineral exploration.* CCDMD.

Nyembo (1997). Etat de connaissance de l'étude petrographique et le bilan minéralogique de la mine de kipushi. *Travail de fin de cycle, Unilu,* 45(12-16).

Olivier, B. J. (December-2005). Study of the babola gold deposit by magnetic prospecting. *école polytechnique de Montréal,* 61(41).

Oosterbosch (1992). Les mineralisations dans le système de roan au katanga. in lambard j. et nicolin p (ed). gisements stratiformes de cuivre en a frique. *Association des services géologiques d'Afrique, Paris,* 282.

Patient, B. V. (2020). Upper Katanga. https://www.caid.cd.